PREHISTORIC ALASKA

ALASKA GEOGRAPHIC
Volume 21, Number 4 / 1994

The Alaska Geographic Society

To teach many more to better know and more wisely use our natural resources...

Editor / **Penny Rennick**

Production Director / **Kathy Doogan**

Staff Writer / **L.J. Campbell**

Circulation/Database Manager / **Vickie Staples**

Marketing Manager / **Pattey Parker**

Postmaster: Send address changes to
ALASKA GEOGRAPHIC®
P.O. Box 93370
Anchorage, Alaska 99509-3370

PRINTED IN U.S.A.

ISBN: 1-56661-024-9

Price to non-members this issue: $19.95

The Library of Congress has cataloged this serial publication as follows:

Alaska Geographic. v.1-
[Anchorage, Alaska Geographic Society] 1972-
v. ill. (part col.). 23 x 31 cm.
Quarterly
Official publication of The Alaska Geographic Society.
Key title: Alaska geographic, ISSN 0361-1353.

1. Alaska—Description and travel—1959-
—Periodicals. I. Alaska Geographic Society.

F901.A266 917.98'04'505 72-92087

Library of Congress 75[79112] MARC-S

COVER: *Dinosaurs roamed Alaska during the Cretaceous. Please see page 29 for identifications. (Painting by Tom Stewart)*

PREVIOUS PAGE: *Windblown volcanic dust partially obscures this view of Mount Griggs in Katmai National Park. (Charlie Crangle)*

FACING PAGE: *Onion Portage in the Kobuk Valley contains a record of prehistoric people stretching back thousands of years. (Chlaus Lotscher)*

ALASKA GEOGRAPHIC® is published quarterly by The Alaska Geographic Society, 639 West International Airport Road, Unit 38, Anchorage, AK 99518. Second-class postage paid at Anchorage, Alaska, and additional mailing offices.

Registered trademark: Alaska Geographic, ISSN 0361-1353; Key title Alaska Geographic.

THE ALASKA GEOGRAPHIC SOCIETY is a non-profit, educational organization dedicated to improving geographic understanding of Alaska and the North, putting geography back in the classroom and exploring new methods of teaching and learning.

SOCIETY MEMBERS receive *ALASKA GEOGRAPHIC®*, a quality publication that devotes each quarterly issue to monographic in-depth coverage of a northern geographic region or resource-oriented subject.

MEMBERSHIP in The Alaska Geographic Society costs $39 per year, $49 to non-U.S. addresses. ($31.20 of the membership fee is for a one-year subscription to *ALASKA GEOGRAPHIC®*.) Order from The Alaska Geographic Society, Box 93370, Anchorage, AK 99509-3370; phone (907) 562-0164, fax (907) 562-0479.

SUBMITTING PHOTOGRAPHS: Please write for a list of upcoming topics or other specific photo needs and a copy of our editorial guidelines. We cannot be responsible for unsolicited submissions. Submissions not accompanied by sufficient postage for return by certified mail will be returned by regular mail.

CHANGE OF ADDRESS: The post office does not automatically forward *ALASKA GEOGRAPHIC®* when you move. To ensure continuous service, please notify us six weeks before moving. Send your new address and your membership number or a mailing label from a recent *ALASKA GEOGRAPHIC®* issue to: The Alaska Geographic Society, Box 93370, Anchorage, AK 99509-3370.

Color Separations / GRAPHIC CHROMATICS

Printed by / THE HART PRESS

ABOUT THIS ISSUE: This project could not have been done without the willingness and diligence of members of the Branch of Alaskan Geology, U.S. Geological Survey in Anchorage. To Dr. Frederic H. (Ric) Wilson, author of the land chapter, and to Dr. David Carter, branch chief; Dr. Tom Hamilton, and numerous other scientists within the branch we say a big thank you. Dr. Tom Ager in the survey's Denver office wrote about prehistoric vegetation, and we thank him and other members of the Denver office for their review of the manuscript. Florence R. Weber, now retired from USGS, co-wrote the lands chapter. For his review of the manuscripts and advice on all aspects of Alaska's geology we thank Dr. Charles (Gil) Mull of the Alaska Geological Survey in Fairbanks, a long-time friend and contributor to the society. We appreciate the assistance of Dr. Robert B. Blodgett of the U.S. Geological Survey in Washington, D.C. in preparing the information for the Geologic Timeline.

Staff writer L.J. Campbell wrote the chapter on dinosaurs, with much thanks to Dr. Roland Gangloff of the University of Alaska Museum, Fairbanks, for his consultation and information. Also we thank Dr. Philip Currie, Royal Tyrrell Museum in Drumheller, Alberta; Dr. Judith Totman Parrish of the University of Arizona, Tucson; Dr. William Clemens of the University of California Berkeley; Anne Pasch of the University of Alaska Anchorage; Dr. Bob King of the federal Bureau of Land Management; Dr. David Hopkins of the University of Alaska Fairbanks; and Dr. David Norton, associate professor of natural sciences at Arctic Sivunmum Iliasaqvik College, Barrow. We thank doctoral student Paul Matheus for his account of Alaska's Pleistocene mammals.

A special thank you goes to Anchorage artist Tom Stewart, who painted the original works on Alaska's dinosaurs and Pleistocene mammals, in consultation with Dr. Gangloff, Dr. Hopkins and Paul Matheus. For their help in obtaining other illustrations we thank Dennis Johnson, Red Deer College Press, Alberta; Marty Hickie, Royal Tyrrell Museum; and Lodvina Mascarenhas, Natural History Museum, London.

Three University of Alaska Anchorage anthropologists contributed articles on prehistoric people: Dr. Bill Workman wrote the overview on Alaska's prehistoric people and reviewed all the manuscripts in this section, Dr. Douglas W. Veltre wrote about prehistoric Aleut culture and Dr. David R. Yesner, with his colleague Kristine J. Crossen, wrote about the Broken Mammoth site and prehistoric people in the Interior. Free-lance writer and former science writer for the *Los Angeles Times* Lee Dye wrote the article on the Mesa site, and we thank Michael Kunz of the Bureau of Land Management for review and update of this material. We thank Jeanne Schaaf, research archaeologist with the National Park Service, and her colleague Herbert Anungazuk for their accounts of prehistoric people on the Seward Peninsula. And we are grateful to Dr. Wallace M. Olson, retired from the University of Alaska Southeast, for his article on prehistoric people in Southeast, and to Dr. Gerald Shields, of the University of Alaska Fairbanks, for his article on molecular evolutionary genetics.

Contents

PRECAMBRIAN ERA

Geologic Timeline

(The ribbon depicts relative duration of Earth's eras to scale, 60 million years per inch. MYA = million years ago. *What is described for this period in Alaska may have formed elsewhere.)

PRECAMBRIAN ERA

5 billion-570 MYA. Spans nearly 4.5 billion years from formation of Earth's crust and oceans to appearance of first complex organisms; see timeline/lavender. Single-celled algae, bacteria appear in seas 3.5 billion years ago. Oxygen atmosphere forms through photosynthesis 2 billion years ago. Multicelled marine worms appear 1,000 MYA. Jellyfish appear 600 MYA.

PALEOZOIC "ANCIENT LIFE" ERA

Most animal, plant groups appear, except flowering plants. See timeline/blue.

Cambrian Period – 570 - 500 MYA

Global: Warm climate. Algae abundant. Marine invertebrates like brachiopods appear. No terrestrial life.

***Alaska:** Warm seas full of trilobites (three-lobed marine arthropods now extinct). Diverse trilobite faunas in Holitna River basin, with strong affinities to Siberian faunas. Cambrian exposures also north of Eagle and in Brooks Range.

Ordovician Period – 500 - 435 MYA

Global: Warm mild climate, warm seas; number of climatic zones increases. Marine invertebrates abundant. First vertebrates are fish. Corals appear.

***Alaska:** Tropical seas with gastropods, trilobites, corals, brachiopods, sponges, calcareous algae. Graptolites, now-extinct branched organisms, floated in deep waters. Ordovician rocks found north of Medfra in Kuskokwim Mountains; in Porcupine and Holitna river basins; along the Brooks Range; near Livengood, eastcentral Alaska; and York Mountains, western Seward Peninsula.

Silurian Period – 435 - 408 MYA

Global: Mild climate. Primitive land plants appear.

***Alaska:** Warm, tropical waters. Limestone deposited in shallow water. Brachiopods, gastropods, corals, trilobites, graptolites, algae in seas. Thick Silurian reefs of blue-green algae widespread, including Southeast (Heceta Limestone on Prince of Wales Island); Shellabarger Pass, west of Mt. McKinley; near White Mountain, south of McGrath; in arc from Lime Village to Holitna River. Similar algal reefs in Ural Mountains, Russia. Silurian non-reef deposits in northern Kuskokwim Mountains; Glacier Bay; Brooks Range; Porcupine River basin; York Mountains.

Devonian Period – 408 - 360 MYA

Global: Moderate climate; some cool water areas. Insects, amphibians appear. Fish develop lungs, move onto land. Vascular plants, like ferns and *Equisetum* (horsetails) appear as do first forests.

***Alaska:** Warm tropical seas. Brachiopods, corals, mollusk groups, including gastropods, abundant. Trilobites less abundant. Volcanoes active locally. Shallow water limestones widespread. Coral reefs develop. Major northern uplift deposits sandstones, conglomerates along area of Brooks Range. Devonian exposures also near Eagle; Porcupine River drainage; Livengood area; Lime Village to Holitna River; and Southeast, including Prince of Wales Island.

Carboniferous Period – 360 - 286 MYA

(Also known as Mississippian, 360 - 328 MYA, and Pennsylvanian, 328 - 286 MYA.)

Global: Lush plant growth in equatorial, mid-latitudes. Gymnosperms, cycads, conifers appear. *Equisetum* grow tree size. Swamp sediments form hard coals. First freshwater mollusks. Amphibians flourish; insects develop wings; reptiles appear.

***Alaska:** Climate warm, but cooling. Warm water limestones still forming; tropical conditions start disappearing. Seas hold brachiopods, corals, mollusks, fusulinids (one-celled, hard-shelled protozoa). Carboniferous limestone in shallow sea forms rocks known as Lisburne Group, widespread in Brooks Range and North Slope. Zinc-lead deposits form in Red Dog area, western Brooks Range. Mississippian fossil plants in central Brooks Range and Pennsylvanian plant fossils near Shellabarger Pass are similar to Siberian flora. Mississippian coal deposits on Lisburne Peninsula, Alaska's oldest coal, indicate heavy vegetation. Seas withdraw from northern Alaska in Pennsylvanian.

Permian Period – 286 - 245 MYA

Global: Cool, dry climate; glaciation in southern hemisphere. Reptiles abundant; mammallike reptiles appear. Conifers abundant; gingkoes appear. Wide-scale extinction at end of Permian takes many shellfish, corals, fishes, reptile groups.

***Alaska:** Cooling climate. Water subtropical in south, temperate in north. Brachiopods, corals, mollusks, bryozoans (lacy organisms) common in seas. Permian sandstone, shale deposited in Brooks Range and near Eagle; volcanic rocks widespread in eastern Alaska Range, Wrangell Mountains, Southeast.

MESOZOIC "MIDDLE LIFE" ERA

"Age of reptiles" as dinosaurs appear, dominate, disappear. Also arrival of some modern mammals, birds, flowering plants, insects. See timeline/green.

Triassic Period – 245 - 208 MYA

Global: Arid to semidry climate. All land connected in giant continent Pangaea. Modern ferns abundant. Tortoises, ichthyosaurs (swimming reptiles), dinosaurs, small mammals appear.

***Alaska:** Shallow north seas hold mollusks, ichthyosaurs. Sandstones, conglomerates form porous sedimentary deposit, future Prudhoe Bay oil reserve. Copper, silver deposits form near Kennicott. Limestones, salt, gypsum, other deposits indicative of aridity deposited in Wrangellia terrane of Southcentral, Southeast; extensive lava flows in same terrane. Fossil of shallow-water ichthyosaur, *Mixosaurus*, found in 1969 near Ketchikan.

Jurassic Period – 208 - 145 MYA

Global: Mild, warm climate; increased rainfall. Pangaea begins separating. Lush vegetation, mostly

ferns, conifers, cycads. Dinosaurs reign. Bony fishes, primitive birds appear.
***Alaska:** Volcanoes erupt along ancestral Alaska Peninsula through southcentral Alaska, into Talkeetna Mountains. Oil-bearing marine shales, silt sandstones deposited in upper Cook Inlet. Ancestral Brooks Range forms as oceanic crust collides with continental rocks. Ammonites (coiled, chambered mollusks) abundant on Alaska Peninsula and in Talkeetna Mountains. On Alaska Peninsula, theropod dinosaur footprints from Jurassic (or Cretaceous) found in 1975 near Chignik; fossil of plesiosaur (near-shore swimming reptile) found in 1922 near Kejulik River.

Cretaceous Period – 145 - 65 MYA

Global: Temperate, warm climate; seasonal cycles. Atlantic Ocean develops. Continents assume familiar positions; Australia still joined to Antarctica. Flowering plants, first modern insects, placental mammals appear. Dinosaurs, ammonites among 75 percent of life forms disappearing at end of Cretaceous, in Earth's second wide-scale extinction.
***Alaska:** Takes more modern shape. Beaufort Sea forms as northern Alaska and ancestral Brooks Range rotate away from Canadian Arctic, collide with southern Alaska to form modern Brooks Range. Arctic Ocean probably ice-free; some sediments suggest seasonal ice. North Slope intermittantly covered by seas, alternating with delta building; abundant coal beds deposited. Crustal fragments collide to form Interior. Gold-bearing rocks emplaced near Nome, Fairbanks, Flat, Iditarod. Seas temporarily cover Yukon, Koyukuk, Kuskokwim basins in mid-Cretaceous; Arctic Ocean linked to Gulf of Mexico by western interior seaway through late Cretaceous. Southcentral swamps form coal-bearing sandstones, future Matanuska coalfield. Ocean trench south of Alaska Peninsula wraps to Southcentral in late Cretaceous; volcanoes active in southern Alaska. Dinosaurs present, based on fossils from North Slope, Talkeetna Mountains; also North Slope fossil finds of sturgeon, advanced bony fish, marsupials, rodent-like mammals, toothed diving bird *Hesperornis*.

CENOZOIC "RECENT LIFE" ERA

Modern plants, mammals flourish; humans appear. See timeline/yellow.

Tertiary Period – 65 - 1.8 MYA

Paleocene/Eocene Epoch – 65 - 37 MYA

Global: Warm, humid climate. Seas withdraw. Most modern plants present. Modern birds and progenitors of modern mammal groups appear, i.e. ungulates, carnivores, probiscideans, primates.
Alaska: Alaska Peninsula probably tropical rain forest. Southeast assumes modern configuration in Paleocene. Juneau area gold deposits form. In late Eocene, Aleutian Islands form; northern Alaska takes present appearance.

Oligocene Epoch – 37 - 23 MYA

Global: Mild temperate to drier climate. Modern grasses appear. Mastodons, rabbits, pigs, modern-looking rodents appear. Extinction of primitive mammals begins.
Alaska: Continental sandstones, conglomerates deposited in upper Cook Inlet, future oil reservoirs. Volcanoes active in Aleutians, Alaska Peninsula.

Miocene Epoch – 23 - 5.2 MYA

Global: Climate cools. Grass plains, grazing mammals increase. Modern forests, apelike primates appear.
Alaska: Significant warming early; later cooling. Herbivorous sea mammal desmostylid inhabited warm, near-shore waters, based on 1994 fossil find at Dutch Harbor, Unalaska. Thick swamp vegetation forms Healy coalfield in Interior and Beluga coalfield, west of Cook Inlet. Deciduous broadleaf and conifer forests reach southern Brooks Range; replaced by conifers during cooler, late Miocene. Interior rivers flow south to Cook Inlet, rather than west to Bering Sea until late Miocene, when Alaska Range, southern coastal mountains rise; volcanoes reactivated in Aleutians; glaciers form in southern mountains.

Pliocene Epoch – 5.2 - 1.8 MYA

Global: Continued continental uplift. Aridity increases, grasslands spread. Modern mammal genera appear, i.e. horse genus *Equus*, which flourishes. Early humans appear near end of Pliocene.
Alaska: Cool, with warm intervals of greater forest diversity. Oil collects in upper Cook Inlet reservoirs. Pliocene deposits near Gulkana Glacier, Alaska Range and near Chicken, eastern Interior. Late Pliocene brings permafrost to Interior; tundralike plants appear. First large-scale glacial event about 2.5 MYA.

Quaternary Period – 1.8 MYA - present

Pleistocene Epoch – 1.8 MYA - 10,000 years ago

Global: Warming, cooling trends. Glacial age with multiple advances, retreats of continental ice sheets; redistribution of plants. Mammoth, bison abundant. Many mammals disappear at end of Pleistocene. Modern humans appear.
Alaska: During glacial intervals, seas drop exposing Bering Land Bridge; expanses of Interior, North Slope, western Alaska remain ice-free, steppe lands with grasses, sedges, herbaceous flowering plants, willow and berry bushes. Mammoths, bison, horses, lions, saber-toothed cats, short-faced bears, saiga antelope, camels among steppe mammals. Patches of ice-free refugia south of Alaska Range, on Kodiak Island and in Southeast. Glaciers grind rocks, soil into fine silt, or loess, windblown into deep deposits, dunes. Humans inhabit Alaska near end of Pleistocene.

Holocene Epoch – 10,000 years ago - present

Global: Glaciers retreat, climate warms. Holocene may be another interglacial episode. Humans affect climate, geology. Species extinctions continue.
Alaska: Placer gold deposits form in stream gravels. Spruce trees, then alders, spread across lowlands from Canada. Coastal forests spread from Pacific Northwest. Modern tundra plants appear.

Sources: *Dinosaur!* (1991); Frederic H. Wilson, USGS, Anchorage; Robert B. Blodgett, USGS, Reston, Va.; Gil Mull, Alaska Geological Survey; William Elder, USGS, Menlo Park, Calif. Timeline graphic by Kathy Doogan.

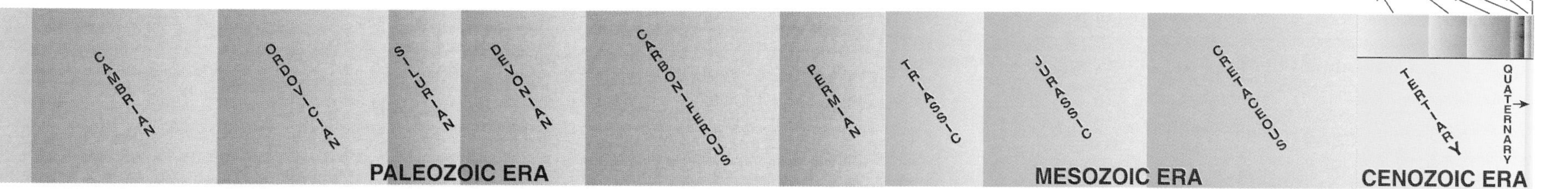

Prehistoric Alaska: The Land

By Dr. Frederic H. (Ric) Wilson and Florence R. Weber

Editor's note: *Dr. Wilson is a geologist with the U.S. Geological Survey in Anchorage. Florence Weber, of Fairbanks, is retired from USGS. She has received an honorary doctorate from University of Alaska Fairbanks. Bold type indicates items in the glossary.*

■ *Introduction*

Many Alaskans know the dynamic nature of Alaska's landscape firsthand. The 1964 earthquake, the 1989 eruption of Mount **Redoubt** volcano, the frequent earthquakes in the Aleutians and the ever-shifting meanders of the Yukon and Kuskokwim rivers remind them of constant changes to the land. These changes are part of the continuing story of the geologic growth and development of Alaska during hundreds of million of years. By geologic **time**, Alaska has only recently come into existence and the dynamic processes that formed it continue to affect it. The landscape we see today has been shaped by glacier and stream erosion or their indirect effects, and to a lesser extent by volcanoes. Most prominently, if less obviously, Alaska has been built by slow movements of the Earth's crust we call **tectonic** or mountain-building.

During 5 billion years of geologic time, the Earth's crust has repeatedly broken apart into plates. These plates have recombined, and have shifted positions relative to each other, to the Earth's rotational axis and to the equator. Large parts of the Earth's crust, including Alaska, have been built and destroyed by tectonic forces. Alaska is a collage of transported and locally formed fragments of crusts. As erosion and deposition reshape the land surface, climatic changes, brought on partly by changing ocean and atmospheric circulation patterns, alter the location and extent of tropical, temperate and arctic environments. We need to understand the results of these processes as they acted upon Alaska to understand the formation of Alaska. Rocks can provide hints of previous environments because they contain traces of ocean floor and lost lands, bits and pieces of ancient history.

■ *The Birth of Alaska*

When did Alaska come into existence? At what time and where do we recognize the first parts of Alaska? This is a difficult question. Alaska is composed of many fragments of geologic **terranes**, each having a wide range of ages, geologic character and site of origin. Geologists have debated the interpretation and origin of these terranes for the last 25 years.

The sequence of events that created Alaska cannot be fully documented. Some rocks of southwestern Alaska are as old as 2 billion years, but geologists aren't sure these rocks formed in North America. Yet, some rocks in interior Alaska 500 million to 1 billion years old were certainly formed in ancestral North America. These rocks show ties to rocks of the Canadian Shield of northcentral Canada, which contains rocks as old as 4 billion years, among the oldest on Earth.

Where continental fragments collide, rocks are folded and faulted. The folding thickens the resulting mass of rocks. Because continental crust is less dense than oceanic crust, it floats on oceanic crust and the mantle. This buoyancy pushes some of the rocks into the air, forming mountains, and pushes others deep into the Earth, where they are metamorphosed. The collision of continental and oceanic crusts results in **subduction**, where the denser oceanic crust is overridden by the less dense continental crust. As this occurs the continental rocks deform, and oceanic rocks may scrape off and plaster against the lower side of the continental crust. This process also builds mountains, the Chugach Mountains in southcentral Alaska for example. At subduction zones, the oceanic crust partially melts, typically at roughly 90 to 95 miles deep, as it dives within the Earth. The resulting magmas rise through weak spots in the Earth's crust, generally interacting with crustal rocks on the way, and finally solidifying as plutonic rocks or erupting as volcanoes. Throughout Alaska, sites of past plate collisions can be recognized by the presence of mountains. Seismically active mountains may indicate ongoing collisions or post-collisional adjustments. Metamorphic rocks generally indicate the sites of ancient eroded mountains, though given the dynamic nature of Alaska's terranes, the mountains themselves could have been located distant from the present location of their remaining roots.

As mountains are being built, erosion continually wears them down. Products of erosion are carried away and deposited, temporarily, in valleys, coastal plains, river deltas and ultimately, the ocean. Erosion and

This stump, 6 feet high, is representative of a petrified forest 15 million to 20 million years old exposed along the northwestern shore of Unga Island in the Shumagins. (Frederic H. Wilson, USGS)

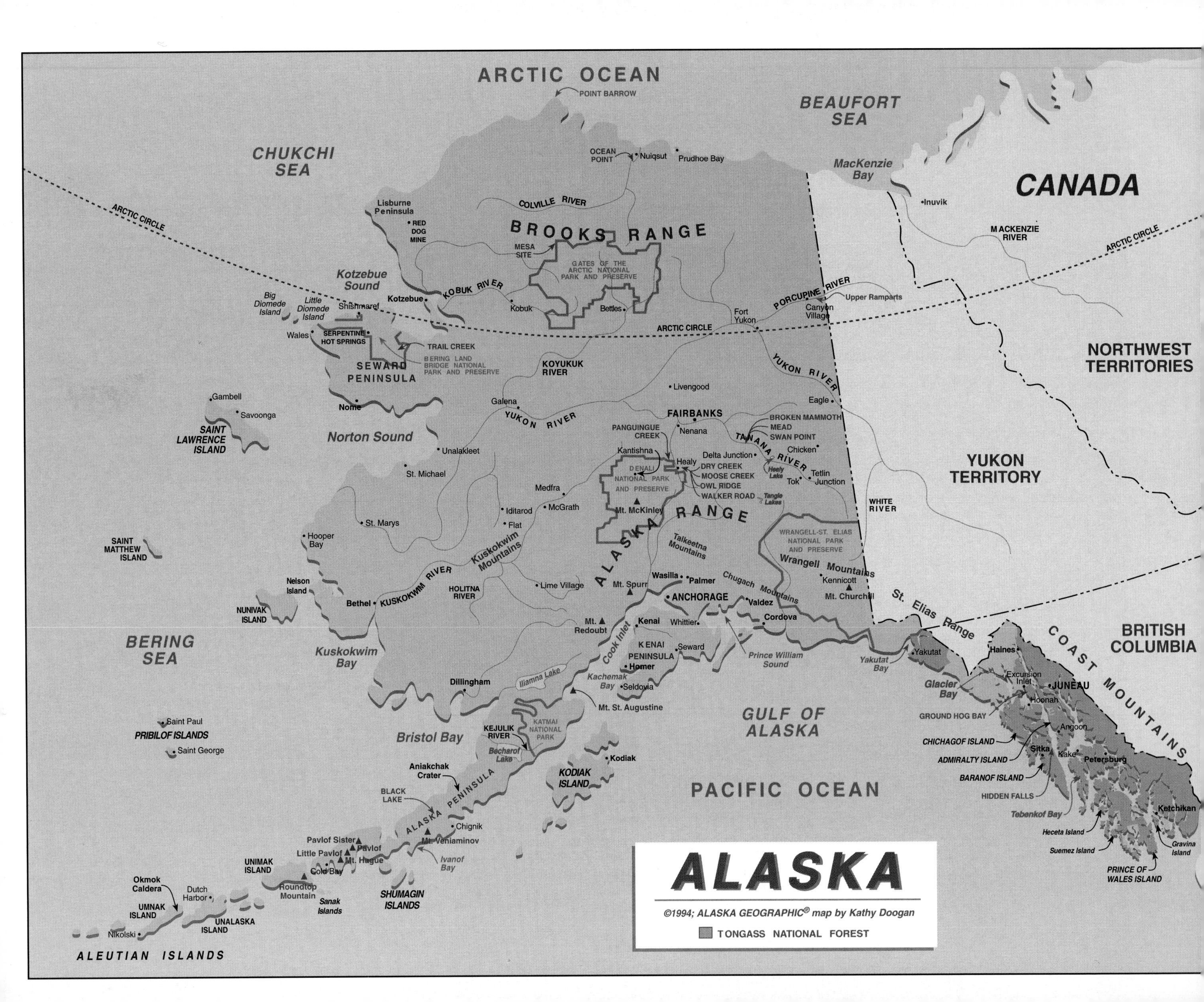

ARCTIC OCEAN
POINT BARROW
BEAUFORT SEA
CHUKCHI SEA
CANADA
OCEAN POINT
Nuiqsut
Prudhoe Bay
MacKenzie Bay
Inuvik
MACKENZIE RIVER
ARCTIC CIRCLE
COLVILLE RIVER
Lisburne Peninsula
RED DOG MINE
BROOKS RANGE
MESA SITE
GATES OF THE ARCTIC NATIONAL PARK AND PRESERVE
Kotzebue Sound
KOBUK RIVER
Kotzebue
Kobuk
Bettles
Big Diomede Island
Little Diomede Island
Shishmaref
Wales
SERPENTINE HOT SPRINGS
TRAIL CREEK
BERING LAND BRIDGE NATIONAL PARK AND PRESERVE
SEWARD PENINSULA
KOYUKUK RIVER
PORCUPINE RIVER
Upper Ramparts
Fort Yukon
Canyon Village
NORTHWEST TERRITORIES
YUKON RIVER
Livengood
Eagle
Gambell
Savoonga
Nome
Galena
SAINT LAWRENCE ISLAND
Norton Sound
FAIRBANKS
PANGUINGUE CREEK
Nenana
BROKEN MAMMOTH
MEAD
SWAN POINT
TANANA RIVER
Chicken
Unalakleet
Kantishna
Delta Junction
Healy
DRY CREEK
MOOSE CREEK
OWL RIDGE
WALKER ROAD
Healy Lake
Tok
Tetlin Junction
Tangle Lakes
YUKON TERRITORY
St. Michael
DENALI NATIONAL PARK AND PRESERVE
Medfra
Mt. McKinley
WHITE RIVER
Iditarod
McGrath
Flat
ALASKA RANGE
St. Marys
SAINT MATTHEW ISLAND
Hooper Bay
Kuskokwim Mountains
Talkeetna Mountains
WRANGELL-ST. ELIAS NATIONAL PARK AND PRESERVE
Wrangell Mountains
KUSKOKWIM RIVER
Nelson Island
HOLITNA RIVER
Lime Village
Wasilla
Palmer
Chugach Mountains
Kennicott
Mt. Spurr
ANCHORAGE
Valdez
Mt. Churchill
NUNIVAK ISLAND
Bethel
St. Elias Range
Mt. Redoubt
Kenai
Whittier
Cordova
BERING SEA
COAST MOUNTAINS
BRITISH COLUMBIA
Cook Inlet
KENAI PENINSULA
Seward
Prince William Sound
Kuskokwim Bay
Homer
Yakutat Bay
Yakutat
Haines
Kachemak Bay
Seldovia
Iliamna Lake
Dillingham
Glacier Bay
Excursion Inlet
JUNEAU
Hoonah
Mt. St. Augustine
GULF OF ALASKA
GROUND HOG BAY
Saint Paul
PRIBILOF ISLANDS
Saint George
KEJULIK RIVER
KATMAI NATIONAL PARK
Bristol Bay
Angoon
CHICHAGOF ISLAND
Becharof Lake
Sitka
Kake
Petersburg
ADMIRALTY ISLAND
Kodiak
KODIAK ISLAND
BARANOF ISLAND
Aniakchak Crater
PACIFIC OCEAN
HIDDEN FALLS
BLACK LAKE
ALASKA PENINSULA
Tebenkof Bay
Ketchikan
Heceta Island
Chignik
Mt. Veniaminov
Pavlof Sister
Pavlof
Suemez Island
Gravina Island
Little Pavlof
Mt. Hague
Ivanof Bay
UNIMAK ISLAND
Cold Bay
PRINCE OF WALES ISLAND
Okmok Caldera
Dutch Harbor
Roundtop Mountain
Sanak Islands
SHUMAGIN ISLANDS
UMNAK ISLAND
UNALASKA ISLAND
Nikolski
ALEUTIAN ISLANDS
ALASKA
©1994; ALASKA GEOGRAPHIC® map by Kathy Doogan
TONGASS NATIONAL FOREST

These rocks show greatly disturbed beds in a mica schist at the leading edge of an accreted terrane on a tributary of the Little Chena River northeast of Fairbanks. (Florence R. Weber)

volcanism create many of the landforms we see today in Alaska.

In the 1960s, Dr. David Stone and his coworkers at the University of Alaska Fairbanks began to measure **paleomagnetic signatures** of rocks in many places in Alaska. They determined that the Alaska Peninsula and other parts of southern Alaska formed far south of their present location. Somewhat tongue-in-cheek, they called this part "Baja Alaska," analogous to Baja California. Stone's idea was originally thought to be pretty farfetched, but now scientists think that it is actually a fairly reasonable description of much of southern Alaska. Additional work by other paleomagnetists, such as Dr. Jack Hillhouse of the USGS, working with paleontologist Dr. David Jones of the USGS, confirmed Stone's hypothesis for a near-equatorial origin for much of southern Alaska. They collected paleomagnetic data from the **Nikolai Greenstone** in the Wrangell Mountains that show these rocks formed at a near-equatorial latitude. Additionally, salt, gypsum and limestone deposits that accompany the Nikolai Greenstone are rock types that form only in intertidal environments at near-equatorial latitudes today. The distinctive stratigraphy they described, in combination with the paleomagnetic evidence, was used to define the Wrangellia terrane, one of the best documented accretionary terranes. However, the great age and complex geologic history of

Construction of Alaska From Its Major Components

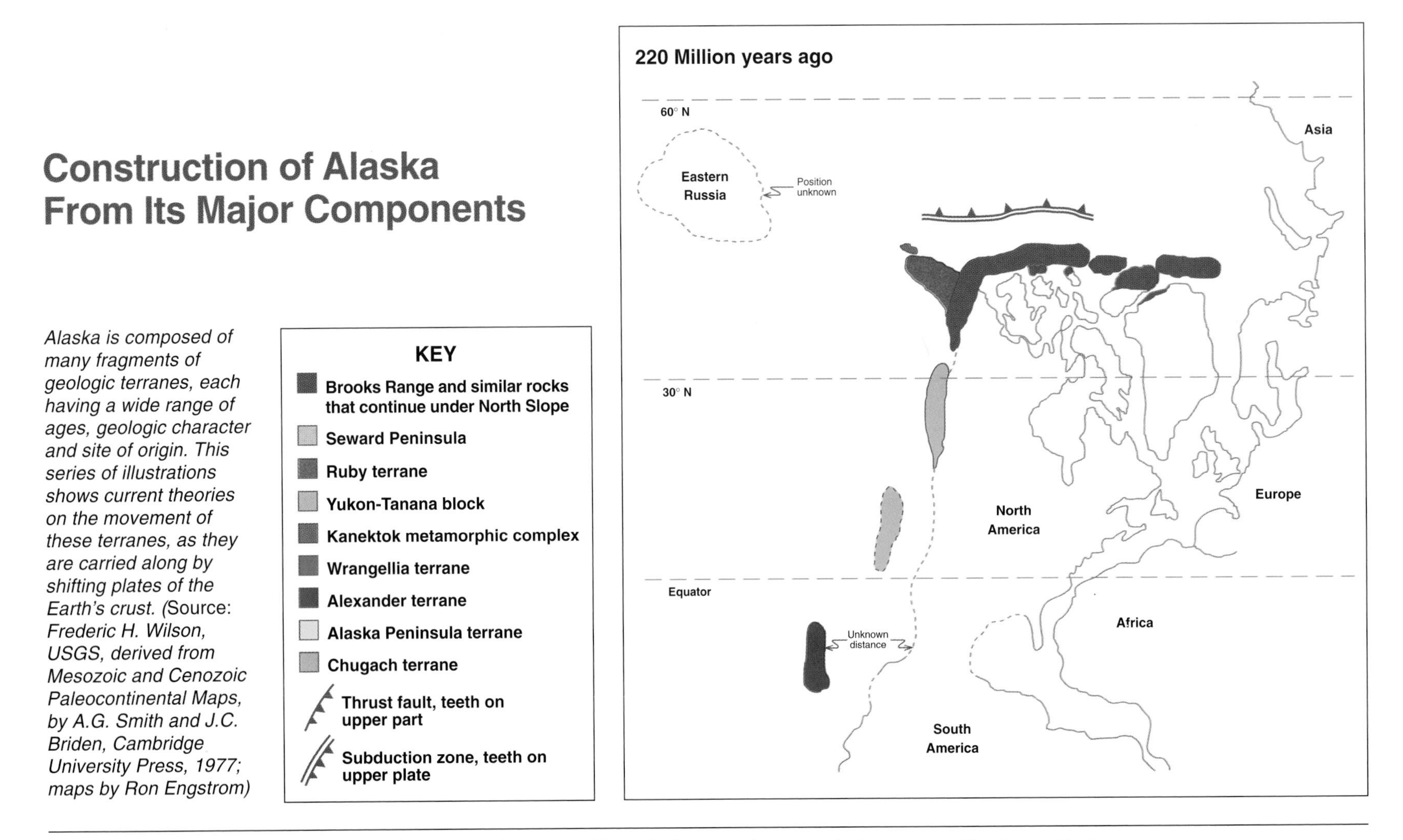

*Alaska is composed of many fragments of geologic terranes, each having a wide range of ages, geologic character and site of origin. This series of illustrations shows current theories on the movement of these terranes, as they are carried along by shifting plates of the Earth's crust. (*Source: *Frederic H. Wilson, USGS, derived from Mesozoic and Cenozoic Paleocontinental Maps, by A.G. Smith and J.C. Briden, Cambridge University Press, 1977; maps by Ron Engstrom)*

the rocks in some terranes often preclude the use of paleomagnetic techniques to determine their latitude of origin. Because of this, the story is not really so simple.

Currently, geologists can divide Alaska into two parts of somewhat different origin. These parts are "North American" Alaska and "Accreted" Alaska. North American Alaska largely consists of those parts formed in North America, although they may be somewhat displaced from their original locations. Accreted Alaska contains many terranes that could be exotic to North America. In a crude sense, the North Slope, Brooks Range and Yukon-Tanana Upland are North American Alaska and southern Alaska is Accreted Alaska. In some areas, the location and nature of the boundary between the two is controversial.

■ *Northern Alaska*

Alaska's North Slope and Brooks Range are largely sedimentary rocks deposited on a continental margin, similar to today's Atlantic coast of North America. In the Brooks Range these sedimentary rocks are deformed and metamorphosed to an increasing degree toward the south. In addition, there are igneous intrusions that are also metamorphosed in the

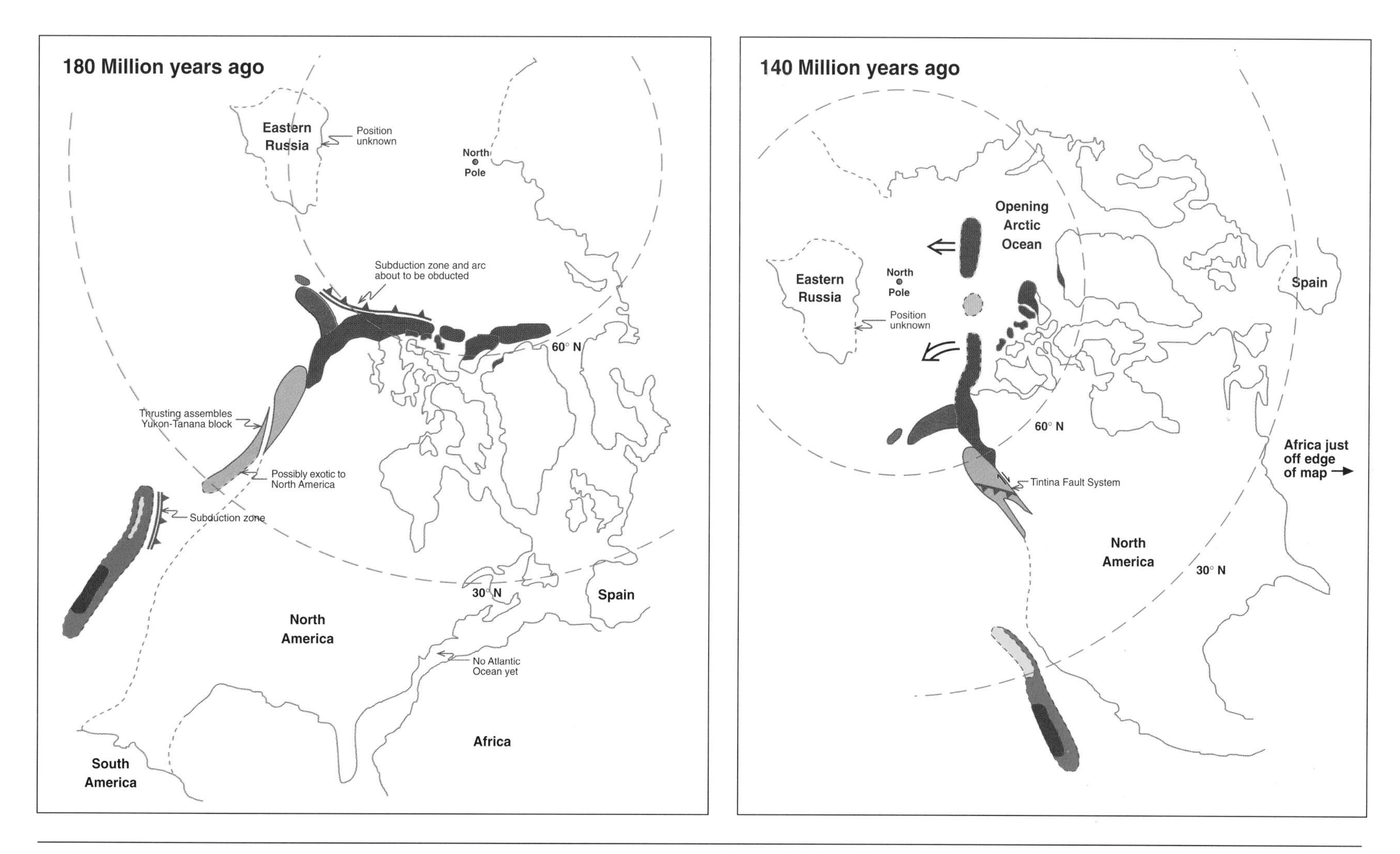

southern part of the Brooks Range. The rocks of northern Alaska are of interest because of vast petroleum accumulations found within them. However, the rocks of greatest interest are buried and only seen in drill cores. Fortunately, the same types of rocks are exposed in the Brooks Range and as a result, the Brooks Range is of great interest.

In northern Alaska, the exposed rocks range back to 1 billion years old, but most of the exposed rocks in the Brooks Range and North Slope consist of Paleozoic and Mesozoic sedimentary rocks or metamorphosed sedimentary rocks that range from 570 million to as young as 50 million years old.

These rocks have long been part of North America, but scientists think that at some time in the past rocks of the Brooks Range and North Slope were elsewhere and may have been oriented in a significantly different direction. Evidence from paleomagnetism and the presence of abundant carbonate rocks such as limestone and dolomite in older rocks in the Brooks Range suggest that these rocks were deposited much farther south than at present. Carbonate rocks are generally deposited at latitudes less than 45 degrees.

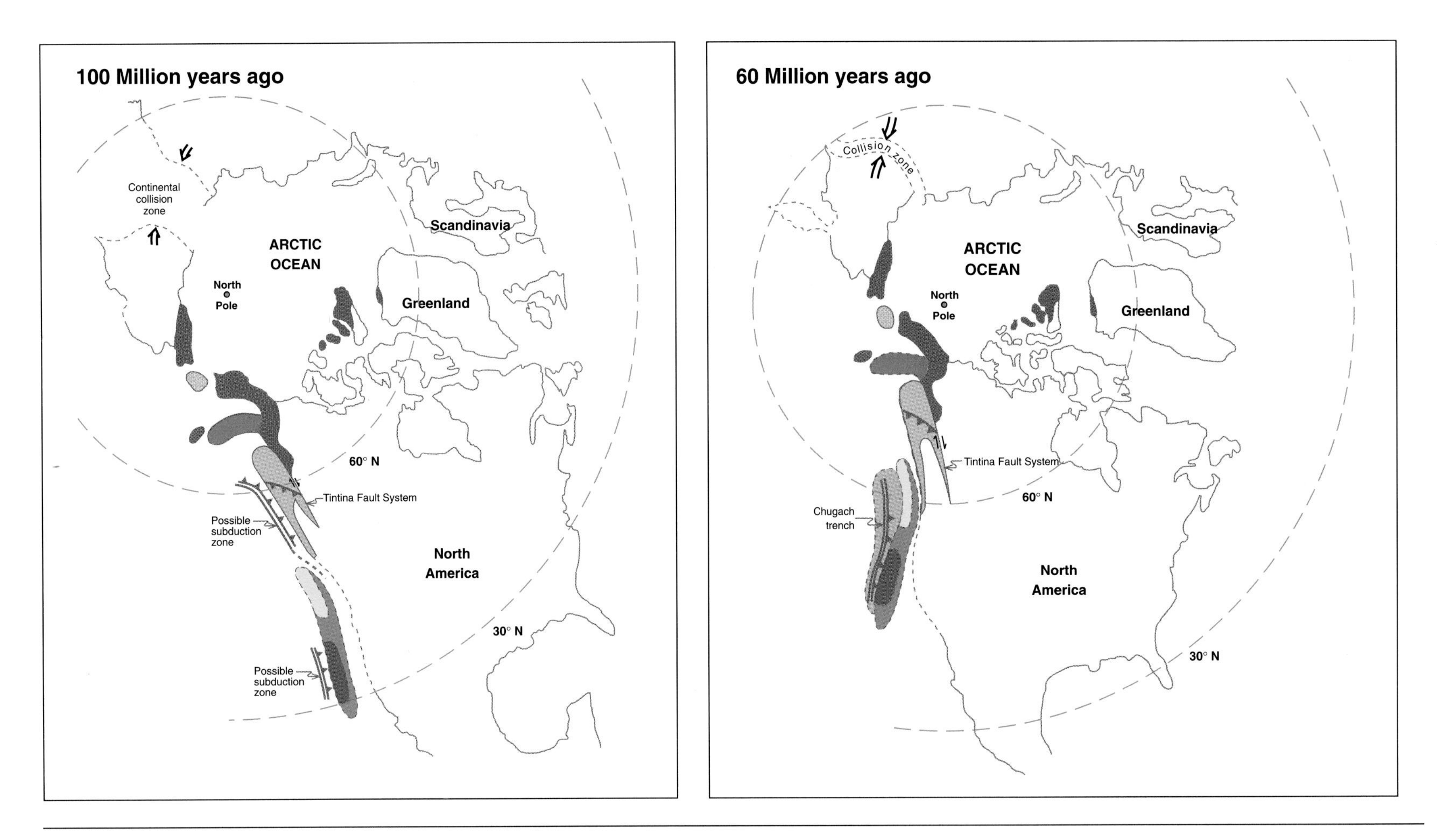

According to geologist Gil Mull of the Alaska Geological Survey, northern Alaska and Canada's Arctic Islands share similar rock units and geologic histories. One current theory for the origin of this region suggests the sedimentary rocks that we see today were deposited on the continental margin adjoining North America in the vicinity of today's Arctic Islands. An oceanic plate adjacent to this margin and consisting largely of igneous rocks such as basalt was being consumed in a subduction zone northwest of the continent. **Magmatic arcs** are commonly associated with subduction zones, and about 160 million to 170 million years ago, the subduction zone and its associated magmatic arc began to collide with the continental margin. Subduction of the continent began; however, the low-density continental rocks were too buoyant to subduct easily and subduction halted. The volcanic arc partially overrode (or was **obducted** onto) the continent. During the collision and obduction, the continental rocks were deformed, partially metamorphosed, and piled up by thrust **faults** to form the core of the ancestral Brooks Range. Their buoyancy and increased thickness resulted in uplift that formed the core of the mountains. Remnants of the obducted volcanic arc remain today as oceanic crustal rocks in a belt along the southern margin of the Brooks Range and in isolated areas in the northern part of the western Brooks Range.

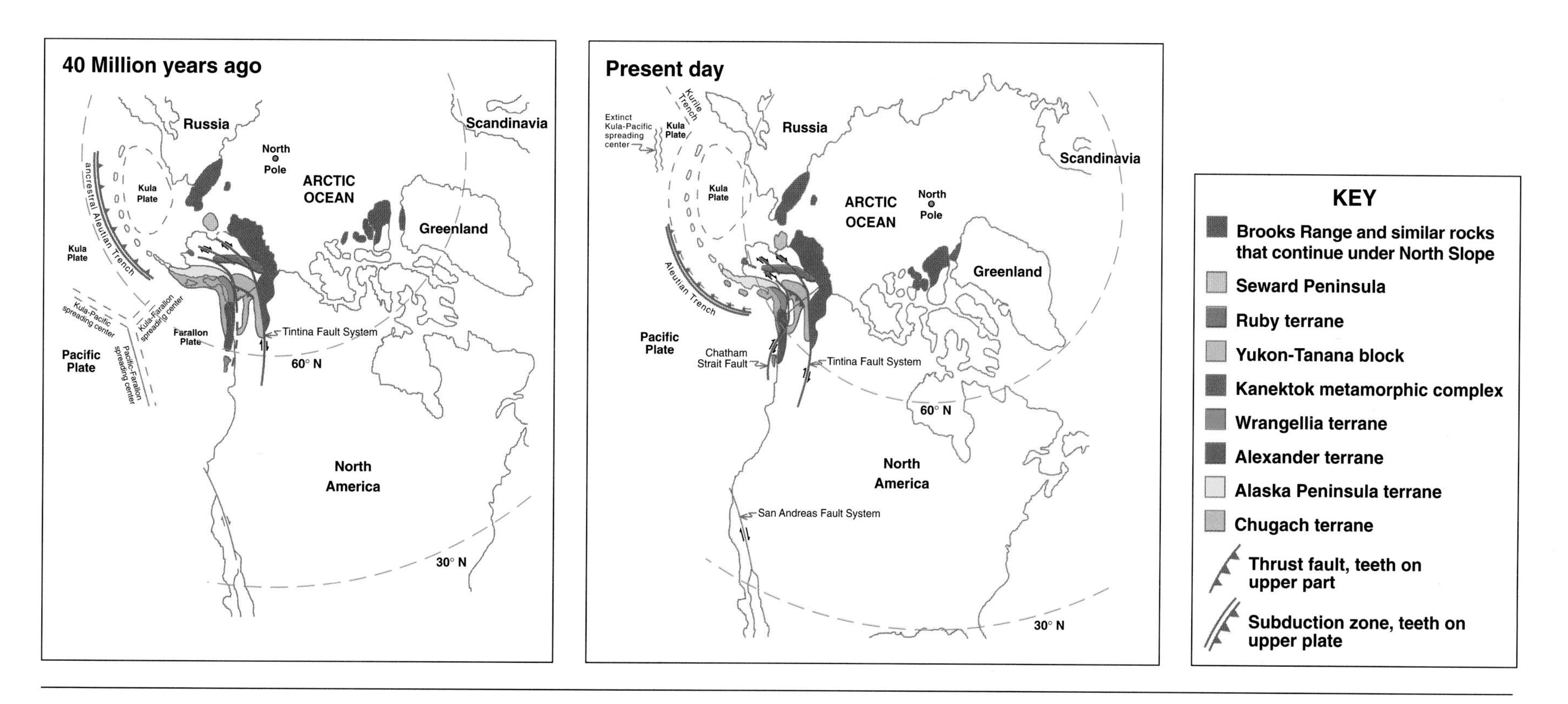

Initially, what is today's North Slope was a shallow sea that lay between the developing Brooks Range and the continental interior of North America, but about 130 million years ago, northern Alaska began to break away (**rift**) from North America. Many geologists and geophysicists think that during millions of years northern Alaska rotated counter-clockwise away from the Arctic Islands about a hinge point that may have been located near the present MacKenzie River delta in the Northwest Territories. This rotation created the Canada Basin of the Arctic Ocean north of Alaska. As the rotation occurred, the Brooks Range continued to rise and shed sediments into the Colville Basin, a smaller basin adjacent to the mountains but separated from the main Arctic Ocean by a ridge or arch located near the present northern Alaska coastline. By about 90 million years ago, much of the Colville Basin was filled and the western part of the North Slope was a low swampy delta that fostered formation of extensive coal deposits. This swampy environment continued to fill the basin, and dinosaurs lived here until about 65 million years ago. This rotation theory remains controversial, however, and geologic **structures** near the hinge point do not fully support this idea.

Long after the compression and rotation ended, erosion became the dominant sculptor of the Brooks Range. Still, as early as 50 million years ago, new tectonic motion began in northeastern Alaska and near the Seward Peninsula in western Alaska. Brooks Range peaks are highest in the east, where tectonic effects are again rebuilding the mountains; westward, elevations decrease. Presently, rocks of the northeastern Brooks Range and adjacent coastal plain are being forced northward and deforming. Geologically young folding and small earthquakes indicate that deformation continues.

The ancient rocks of the Seward Peninsula encroach on Alaska from the west. Similar to rocks in Russia's Far East, the rocks of the Seward Peninsula also share some geology with the Brooks Range. Because of North America's northward and westward tectonic movement, North America and its accretionary terranes have collided with Asia. This collision may be responsible for formation of the Verkhoyansk Mountains in northeast Russia. and may also force the Seward Peninsula against western Alaska.

■ *Interior Alaska*

The Yukon-Tanana Upland in eastcentral Alaska is a large block of metamorphic rocks that extends southeastward into Canada's Yukon Territory. Consisting of several ancient terranes assembled prior to 160 million years ago, parts of the upland can be related to North America by rocks showing ties to the Canadian Shield. These ancient terranes were metamorphosed between 200 million and 120 million years ago and include some of the oldest rocks in Alaska, having ages 500 million to 1 billion years old or more. Yet, some blocks or terranes of the upland have characteristics that are not clearly North American. These blocks may be younger, are perhaps thrust over the older rocks, and may have been accreted to North America. Other rocks may be oceanic crust obducted onto the continent. The metamorphic rocks of the Yukon-Tanana Upland have also been intruded by numerous granitic bodies roughly 60 million, 90 million to 100 million, and 180 million years old.

A number of strike-slip fault systems, similar to the San Andreas system of California, cut across interior and southern Alaska. Possibly the oldest is the Tintina fault system, which the Alaska Highway follows in the eastern Yukon. The Yukon River also roughly follows the fault system in the westernmost Yukon near Dawson City and in eastern Alaska. Scientists think the fault system was first active about 160 million years ago and that rocks on the southwest side of the system have traveled northwestward between 280 and 435 miles relative to North America. The fault system probably originally extended northwestward across ancestral Alaska. Beginning about 130 million years ago, the system was deflected and curved southwestward as northern Alaska rifted and began to rotate. Metamorphism 120 million years ago and granitic intrusions 60 million, and 90 million to 110 million years ago in the Yukon-Tanana Upland probably reflect episodes of motion along the fault system. Today, the Tintina system continues to account for a small part of the relative motion between portions of southern and northern Alaska.

The Denali fault system runs south of and nearly parallel to the Tintina fault system. Forming a large arc extending from west of Mount McKinley, east through Alaska and the Yukon, the Denali system continues past Kluane Lake in Yukon Territory. The fault system eventually joins the Chatham Strait fault extending northward from southeastern Alaska. Readily visible on topographic maps, the Denali system forms a trough along the core of the central Alaska Range. A sharp bend in the fault system occurs in the Denali National Park area and the rocks are consequently deformed there. One result of the deformation is the high mountain mass of which Mount McKinley (20,320 feet) is part. In the last 50 million to 70 million years, rocks south of the fault system have moved at least 250 miles northwest and west. Frequent earthquakes along the fault system show the motion continues and that the mountains are growing.

Faulting has offset these coal beds near Healy. (Florence R. Weber)

Along the trans-Alaska pipeline corridor, between the Brooks Range and north of the Yukon River, a series of granitic plutons intrude the core of northeast-to-southwest-trending metamorphic rocks called the Ruby terrane. The Ruby terrane may have a tectonic history similar to rocks of the Yukon-Tanana Upland. This terrane also includes rocks metamorphosed as much as 1 billion years ago. Splays of the Tintina fault system cut across the Ruby terrane, and near Galena a portion of the Ruby terrane has moved southwestward at least 80 miles. On the northwest side of the terrane, oceanic crustal rocks that underlie the Yukon-Koyukuk Lowland have been obducted over the Ruby terrane, similar to the volcanic arc that overrode the southern Brooks Range.

Although many of the terranes of Alaska contain old rocks, the oldest known rocks are found in southwestern Alaska; they are more than 2 billion years old. Called the Kanektok metamorphic complex and the Idono Complex, these rock groups are largely metamorphic rocks derived from a continental plate surrounded by much younger rocks. [**Editor's note:** *Names formally designated by the USGS are capitalized, i.e. Idono Complex; those not formally designated are not capitalized.*] Of controversial origin, it is unclear if these rocks formed in North America or were rifted from another continent. The structures in these terranes and the surrounding rocks imply that they are slivers incorporated from elsewhere. How can these slivers be incorporated? Looking at what is happening in California will help us to understand Alaska. The Pacific plate moves north-northwestward past California about 3.5 to 4.5 inches per year. At the same time, North America moves westward, away from the Mid-Atlantic Ridge. As North America impinges on the Pacific plate, faults such as the San Andreas system carry pieces separated from North America northwestward as part of the Pacific plate. For example, Baja California, the Catalina Islands and parts of west coastal California are on their way north now. Depending on the motion of the Pacific plate, some of these fragments may collide with Alaska and some may eventually collide with Asia.

Among the oldest rocks yet found in Alaska are those of the Kanektok metamorphic complex north of Thumb Mountain in the Goodnews Bay area. This photo shows folded metamorphic rocks (migmatite) from that complex. (Frederic H. Wilson, USGS)

■ Southern and Southeast Alaska

The Alaska-Aleutian Range cutting across Alaska in a great east-west trending arc is an important feature of southcentral Alaska. Still forming and growing, these are the highest mountains in North America and include Mount McKinley. Elevations are highest in Southcentral and gradually decrease to the southwest, where the volcanoes of the Aleutian volcanic arc cap the mountains of the range, adding significantly to their height. The arc is being built by the collision of crustal blocks moving northward along the North American margin with blocks already there. Among the blocks in the Alaska-Aleutian Range, a number may be exotic to North America and include captured ocean floor, volcanic island arcs and continental fragments. These fragments give southern Alaska its accretionary character.

Southern Alaska has been likened to a

collage because of the diversity of its geologic blocks or terranes. Shortly after Stone's pioneering paleomagnetic studies, the terrane concept was employed to assign the rock units of Alaska, and particularly southern Alaska, into terranes. In a flurry of activity in the 1970s and early 1980s, geologists defined terranes and terrane boundaries throughout Alaska. The defined terranes ranged from large blocks encompassing hundreds of square miles to small slivers of only a few square miles. A variety of terrane maps exist now, each map slightly different depending on terrane definitions. In the 1990s, better geologic knowledge has led to combining many terranes.

The Peninsular terrane, essentially the same as Stone's "Baja Alaska," was originally defined based on similarities between southcentral Alaska and the Alaska Peninsula. Later work has shown that the concept of Baja Alaska and the definition of the Peninsular terrane were relatively loose interpretations. Nevertheless, Peninsular terrane lore has become a key component of many models describing southern Alaska. Recent research and better-described rocks tie the Alaska Peninsula and its geologic history to the Wrangellia terrane. The emerging picture suggests that much of southern Alaska moved as a large composite block of terranes. Internal faults have rearranged components and sometimes obscure geologic ties between some components.

Geologically, southeastern Alaska and northwestern British Columbia have the character of a way station for the terranes or blocks moving north. Like most way stations, bits and pieces of "junk" get left around. The bits and pieces of geology left in southeastern Alaska tend to be highly deformed relics of some of the same rocks seen elsewhere in southern Alaska. Thus, bits of the Wrangellia, Chugach and possibly even the Alaska Peninsula terranes can be identified in southeastern Alaska and coastal British Columbia.

The Alexander terrane, forming much of southeastern Alaska, might be passing through the way station now. It is composed of carbonate and volcanic rocks roughly 220 million to 600 million years old that were deposited on an oceanic platform. Some researchers, such as Dr. Susan Karl of the USGS, think the ancient rocks hidden under the Alexander terrane may be a small continental block formed independently of North America or may be a piece of North America rifted away prior to 600 million years ago. In either case, the terrane was far enough from a continent

Major Terranes of Alaska

*This map shows the generalized locations of the major terranes or geologic blocks and fault systems in Alaska. (**Source:** Frederic H. Wilson, USGS; graphic by Kathy Doogan)*

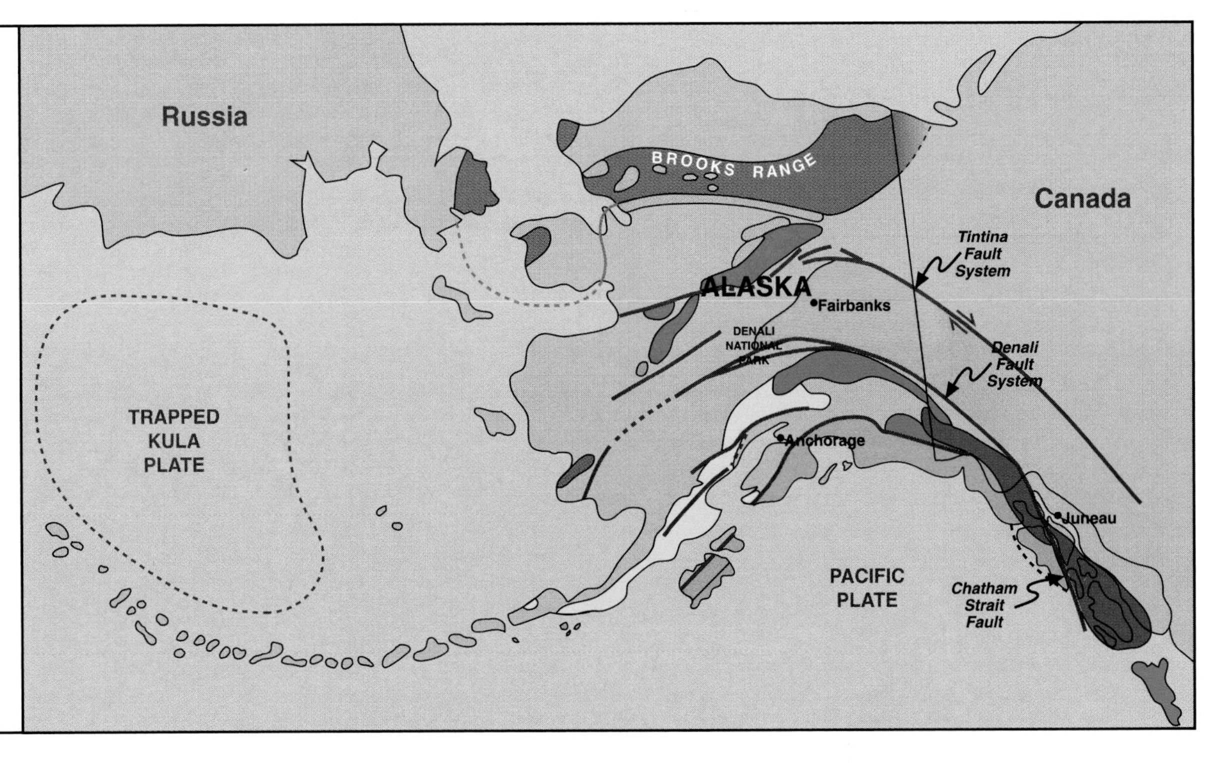

that none of its rocks were derived from a continental source. About 220 million years ago, the Alexander terrane began to rift apart. The process stopped before parts of the terrane were fully separated and the rocks that record this event form the Wrangellia terrane. Because of this direct tie between the Wrangellia and Alexander terranes, strictly speaking, they are the same terrane. The ties between Wrangellia and the Alaska Peninsula terrane have led some researchers to speak of the Peninsular-Alexander-Wrangellia composite terrane. However the composite terrane originally formed, it apparently collided with North America in southeastern Alaska between 150 million and 100 million years ago. After collision, parts continued to move northward. Also, about 100 million years ago, subduction of an oceanic plate beneath southeastern Alaska formed a magmatic arc on the continental margin. The arc's volcanic rocks have eroded away, but its granitic root cuts across the Alexander terrane. Later, 50 million to 65 million years ago, renewed subduction between the Pacific and North American plates created another volcanic arc slightly farther inland. This arc has also eroded and its voluminous granitic root, known as the Coast plutonic complex, is exposed along the boundary between southeastern Alaska and British Columbia in the Coast Range.

Inland from the Alexander terrane are other terranes or fragments of terranes. Less obvious in their relationship with other blocks and in their geologic history due to intense metamorphism, it appears that these rocks have also been transported along the North American margin, some possibly great distances. Another set of terrane terminology, complementary to the models used for Alaska, is applied to these terranes.

The Chugach Mountains in southcentral Alaska and the St. Elias Range in southern Yukon and eastern Alaska are part of the coastal ranges, the youngest of Alaska's mountains. These mountains are largely made of uplifted rocks that formed the continental edge and the fill of the oceanic trench at a subduction zone. Sea-floor sedimentary and volcanic rocks were scraped off the descending oceanic plate by the snowplow effect of the overriding continental plate, as were sea mounts and fragments of other crustal rocks carried along on the oceanic plate. The snowplowing of rocks off the oceanic plate highly deforms (folds) them and the resulting rock is called a **melange**. At the same time the melange was forming, a deposit of sandstone and shale, called **flysch**, was forming in and near the trench. The melange and flysch of southern Alaska are together called the Chugach terrane, one of the most continuous terranes in Alaska. It is exposed from Chichagof Island in southeastern Alaska, around the Gulf of Alaska, and southwest to the Sanak Islands off the Alaska Peninsula.

Solifluction lobes create scalloped patterns on the hillside behind John Cady, who holds a portable magnetometer for geologic field work near the Prindle volcano in eastcentral Interior. (Florence R. Weber)

■ *Alaska Peninsula and the Aleutian Islands*

The Alaska Peninsula and the Aleutian Islands form a large arc across the North Pacific and are well-known for the volcanoes that extend along them. Yet, the widely distributed 100-million-to-200 million-year-old

The floodplain of the Tanana River near Cathedral Bluffs exposes a modern-day source of loess, wind-blown silt created when glaciers scoured the landscape during the ice ages. Loess blankets much of the Interior and has rounded some of the region's landforms. (Frederic H. Wilson, USGS)

sedimentary rocks that form the peninsula's core are less well-known. These rocks record the eruption and erosion of a magmatic arc active between 155 million and 195 million years ago.

About 195 million years ago, the Alaska Peninsula was a chain of volcanic islands. Erosion of the volcanoes deposited conglomerate, sandstone and siltstone on a shallow-water shelf. As the relief decreased, the sediments became finer-grained because rivers decreased in their ability to carry large sediment loads. About 100 million years ago, a fossil-rich sandstone was deposited on a beach with high wave action. Rocks between 70 million and 100 million years old are missing, possibly eroded away. About 70 million years ago, conglomerate and sandstone were derived from erosion of older sedimentary rocks. These coarser rocks become more fine-grained and flyschlike toward the Shumagin Islands where similar-aged deposits are the flysch of the Chugach terrane. Erosion and possibly minor volcanism on the Alaska Peninsula continued until formation of the Aleutian volcanic arc.

For many geologists, the Aleutian Islands offer a classic example of the subduction of one oceanic plate under another. Roughly 40 million years ago, a plate named the **Kula plate** ruptured and one part began to subduct under the other. Well-accepted theory ties volcanism to subduction. The start of volcanism on the Alaska Peninsula can be reasonably linked to the initiation of subduction at the Aleutian Trench, which adjoins both the Alaska Peninsula and Aleutian Islands. On the Alaska Peninsula, the beginning of volcanism was 42 million years ago. So what may have caused the Aleutian Islands to begin to form? Earlier, we talked about the transport and collision of many different terranes to form southern Alaska. As these terranes move and reach a subduction zone, because they are too buoyant to subduct, they tend to plug subduction zones. In some cases, because tectonic motion continues, the subduction zone reforms behind the accreted terrane. The terranes forming southern Alaska have continued to move after collision, typically westward, for at least the last 30 million to 50 million years. Because the shift is longitudinal, it is not directly measurable by paleomagnetic techniques. Scientists speculate that these terrane fragments overrode the oceanic plate being subducted in the Bering Sea Basin. This could have created an instability, rupturing the oceanic plate, trapping part of the Kula plate in the Bering Sea, and forming the Aleutian arc.

The Pacific plate dives under the Alaska Peninsula at a right angle to its trend near Kodiak. Yet, at the longitude of Attu Island in the western Aleutians, the plate moves parallel to the island arc. As a result, no subduction takes place at Attu; the strike-slip motion may actually stretch out and smear the rocks at Attu along the arc. Today, Buldir Island in the western Aleutians is apparently the last place there is enough subduction to generate volcanism. As recently as 10 million to 15 million years ago, volcanoes were distributed along the full length of the Aleutian Islands, though less on the Alaska Peninsula, suggesting more direct convergence in the west.

■ *Climate*

Up to this point, climate has been used to describe the environment in which rocks, such as the carbonates of the Wrangellia terrane, were deposited. In this instance, the environmental information supports the paleomagnetic interpretation that these rocks formed near the equator. But we also have examples where climate has varied greatly throughout time in places of no latitudinal shift. Rocks on Unga Island in the Shumagins contain the remains of a petrified forest about 20 million years old. In this forest Metasequoia grew, an ancestor to giant *Sequoia* trees that today grow in California. According to Dr. Thomas Ager of the USGS, the forest also contained oaks, elms and walnuts, suggesting a climate similar to Pennsylvania or New York today. Mean annual temperature in the forest was about 48 degrees based on the plant community. Today, the region has a mean annual temperature of 40.7 degrees, and trees only grow there if planted from seedlings and sheltered; they do not reproduce. Paleomagnetic studies indicate the rocks on Unga Island formed near where they are now, leaving a changing climate as the best explanation for the difference. Older, 50-million-to-60-million-year-old rocks near Ivanof Bay on the Alaska Peninsula, not far from Unga, also yield plant and animal fossils. The plant fossil evidence suggests a hot rain forest environment having a mean annual temperature of 71.6 degrees even though paleomagnetic evidence suggests Ivanof Bay's rocks formed near their present latitude. At the other extreme, all of these areas lay beneath an ice sheet about 12,000 years ago.

Studies of past climate are an important consideration in reconstructing Earth's history. Our assumption that carbonate rocks only form at low latitude is based on what we see today. Yet, the Ivanof Bay example shows climates could have been significantly different in the past, meaning some assumptions could be invalid. For example, some data indicate that the Earth has been significantly warmer in the past and that the last few million years have been one of the coolest times in Earth's history.

■ *Glaciation and Permafrost*

Today, more than 95 percent of the glaciers in the United States occur in Alaska, and most of these are in southern and southeastern Alaska. The Alaska Range is heavily glaciated and glaciers still play an important role in modifying its landforms. Northern Alaska is certainly cold enough and a few glaciers remain in the eastern Brooks Range, yet in general there is insufficient moisture to build and maintain glaciers. Glaciation, however, is

Stone polygons mark this plateau in the Yukon-Tanana Upland. A periglacial feature, these polygons are of uncertain origin but are related to sorting by frost. (Florence R. Weber)

responsible for many of the landforms of the Brooks Range.

About 2.5 million years ago, glaciers began to accumulate throughout much of the northern hemisphere. The last major episode of glaciation ended about 10,000 years ago. Alaska did not escape this climatic change, yet in a number of ways, the effects upon Alaska were not what most would expect. In the contiguous United States, multiple glacial advances have occurred in the past 2.5 million years; the record is not as clear in Alaska. However, during each of these major worldwide glacial advances, the Bering Land Bridge between Alaska and Asia was exposed as sea level dropped due to the vast amount of water trapped in glacial ice. When glaciers were at their maximum extent, sea level was as much as 400 feet lower than today. Large continental ice masses covered Canada and the northern United States much like Greenland today. In contrast, only alpine glaciers and ice fields developed in Alaska. There were ice fields in the Brooks and Alaska ranges, and in the coastal mountains from southeastern Alaska to the Alaska Peninsula and eastern Aleutians. Smaller alpine glaciers formed in the Yukon-Tanana Upland, higher elevations of the western Aleutian Islands, on the Seward Peninsula and in other isolated parts of interior Alaska.

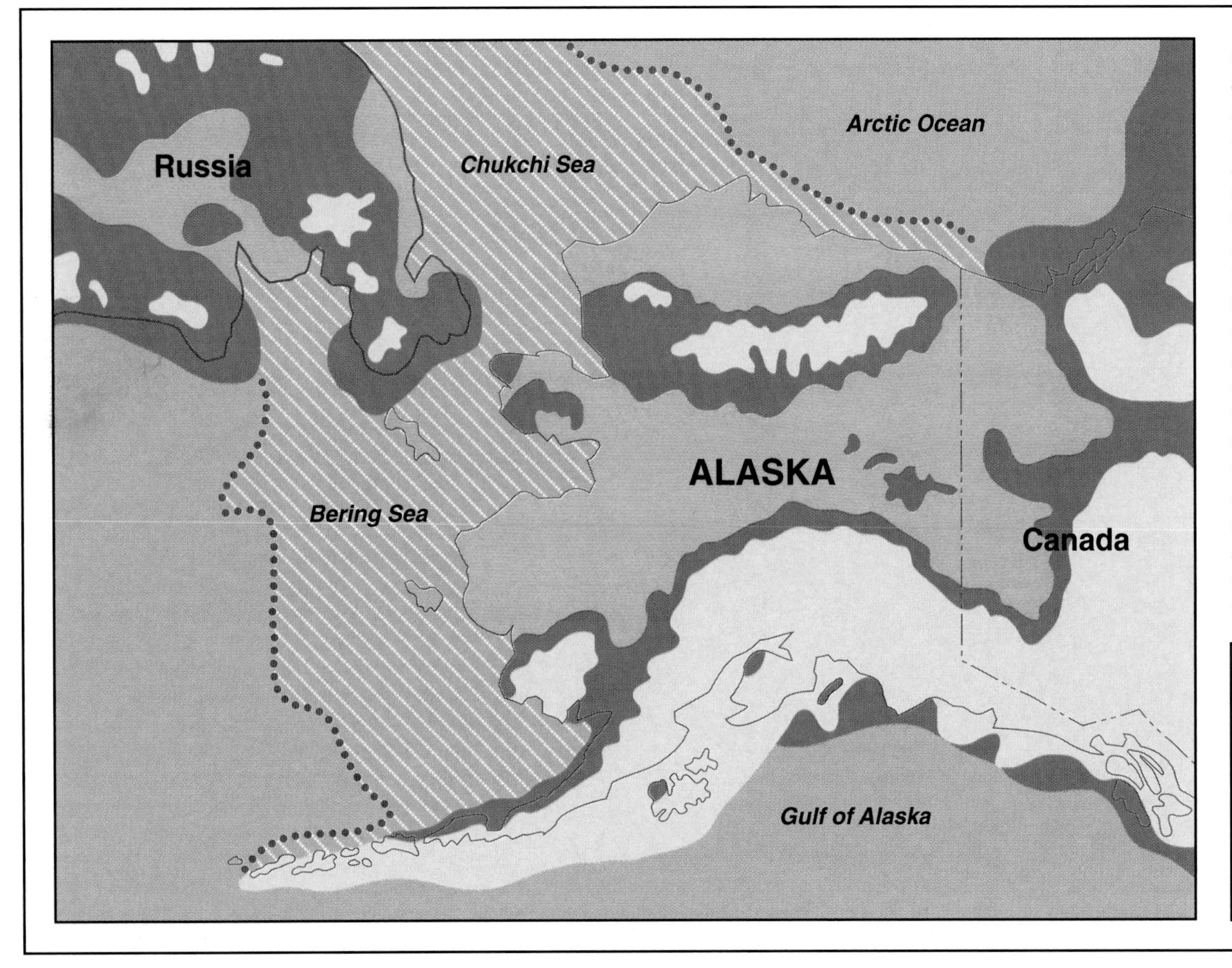

Extent of Glaciation. *This map of Alaska and adjacent areas shows the maximum known extent of glacial cover during the late Pliocene and Quaternary. During numerous intervals of cold global climates the past 2.5 million years, glaciers have covered about a third to a half of the present area of Alaska. Interior, western and northernmost Alaska were too moisture-starved to support creation of large glaciers. In southern Alaska, the development of vast glacier systems resulted in lowering sea level 300 feet or more, exposing large areas of the continental shelf as dry land north and west of Alaska. This exposed land formed a broad bridge up to 620 miles (north to south) that permitted the exchange of plants, animals and humans between northeastern Asia and North America. (***Source:** Glaciation in Alaska, *ed. Thomas D. Hamilton et al., 1986; courtesy of Thomas Ager. Graphic by Kathy Doogan)*

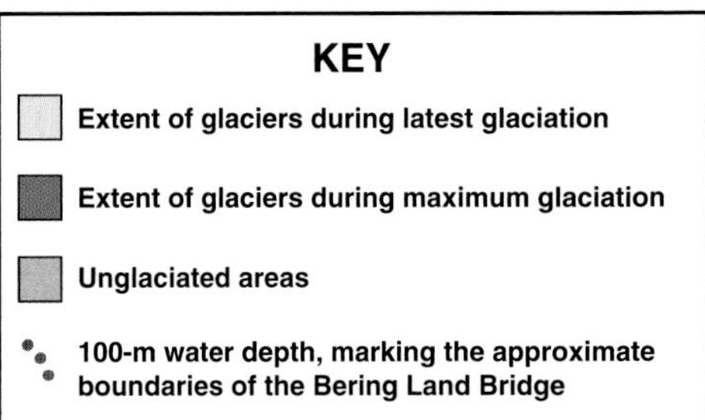

But large parts of interior Alaska and the Bering Land Bridge remained ice free. In fact, on the basis of many biological and geological studies, the interior was virtually a northern Serengeti, having abundant large mammal wildlife, most of which is now extinct. The exposed Bering Land Bridge provided a migration path for plants and animals, eventually including humans, to move back and forth between Asia and North America.

The accelerated erosion caused by glaciers deposited large amounts of sand, gravel and silt in the Kobuk, Yukon, Kuskokwim and Tanana valleys. The North Slope and the Alaska Peninsula's western coastal plain also have extensive glacial deposits derived from adjacent mountains. The mountains of the Yukon-Tanana Upland, although not eroded by the big glaciers, nonetheless do show effects of glaciation. As large deposits of glacial debris accumulated in the Tanana Valley from Alaska Range sources, dry conditions allowed winds to pick up silt and blow it into the hills to the north. The resulting deposits, called "loess", blanket and round the shapes of the hills, sometimes to depths of as much as 200 feet. The deposits provide soils for vegetation. Throughout time, the loess has washed down into valleys where it covers and preserves fossil remains and archaeological sites. In other places such as at Tetlin Junction, Kantishna, in the Koyukuk Lowlands, in the Kobuk River valley and on the North Slope, sand dunes developed. Today, these dune fields are largely stabilized and vegetated, though they preserve a record of a drier, windier time.

Since the rapid retreat of glaciers from their maximum about 10,000 years ago, there have been a number of minor episodes of glacial advance and retreat. Four hundred to 500 years ago, glaciers around the world started advancing during an episode called the Little Ice Age that lasted until about 1900. Many glaciers in southern Alaska advanced and were far more extensive then than they are today, although they did not approach the previous maximums. Today, most glaciers in Alaska are receding, apparently due to climatic warming. Yet in some areas, a few glaciers are advancing or stable. We still have much to learn about how glaciers respond to climate.

Folded beds depict Earth's movement at Calico Bluffs along the Yukon River near Eagle. (Florence R. Weber)

Permafrost can exert major control over the character of the landscape. Many features and landforms develop in permafrost or **periglacial** environments, ranging from pingos, patterned ground and thaw lakes, to ice wedges and ice lenses. In areas of low relief, frozen ground prevents drainage of precipitation, creating swampy ground and many small lakes. On the North Slope, prevailing wind direction causes selective thawing and erosion that create elongated, subrectangular lakes oriented at right angles to the prevailing wind.

Disturbance of ice-rich permafrost produces slumping, ground cracks and other changes. When the activities of man cause these disturbances, major engineering headaches result. In other cases, thawing of ice-poor permafrost, such as bedrock, generates few engineering problems. Though uncommon, some relict permafrost exists in Anchorage. As one travels northward or to higher elevations, permafrost becomes more common. Near Fairbanks, ice-rich permafrost is a major concern because small impacts can disturb it and cause melting. Many structures in interior Alaska show damage or have been destroyed by improper construction techniques in a region of permafrost. North of the Brooks Range, the ground is so solidly frozen that slight warming

LEFT: *Sand dunes are creeping west-southwestward in the Kobuk Valley. The two major dunes are the Great Kobuk, with 25 square miles of active dunes, and Little Kobuk covering 5-plus square miles. From their lowest point next to the Kobuk River southward to their highest point, dunes in the Great Kobuk field rise 275 feet. Much smaller dunes covering about 1 square mile are located on the south side of the Kobuk River across from the mouth of the Hunt River. These dunes were originally part of one large dune field formed during the Pleistocene from loess blown off the Brooks Range and from silts carried by the rivers and left behind when the rivers changed course. (Gil Mull)*

ABOVE: *Within Koyukuk National Wildlife Refuge lies 5-mile-wide Nogahabara Sand Dunes northwest of Roundabout Mountain, an approximately 10,000-acre active dune field. (Florence R. Weber)*

is less of a problem, although in areas of ice-rich permafrost, significant landform changes occur with thawing.

■ ***Volcanoes***

Volcanism is just as important today in Alaska as it has been in the past with numerous active volcanoes concentrated on the Alaska Peninsula, the Aleutian Islands and in the Wrangell Mountains. Lava and volcanic debris flows sometimes fill the valleys that glaciers and rivers have cut, in some cases melting through glaciers. Massive eruptions at Okmok **Caldera** on Umnak Island, Emmons Caldera near Cold Bay, Mount Veniaminov, Aniakchak Crater and others have produced **tephra** deposits that blanket large areas. Such large eruptions also cause short-term worldwide climate fluctuations. A geologically important ash deposit recognized in many areas of the state, the Old Crow tephra, may have been derived about 140,000 years ago from the eruption of the volcano that formed Emmons Caldera. In the last 10,000 years, the volcanoes of Pavlof, Pavlof Sister, Little Pavlof, Mount Hague and Double Top have grown on the rim of the caldera. Twice, roughly 1,200 and 1,800 years ago, a volcano in the Wrangell Mountains, possibly Mount Churchill, erupted, depositing the White River Ash Bed over eastern Alaska and a large part of the Yukon. Today, this ash is seen by many travelers as a white deposit exposed in roadside ditches between Tok and Whitehorse along the Alaska Highway. In 1912 a volcanic eruption disrupted commerce and transportation in Kodiak and led to creation of what is now Katmai National Park. A similar eruption today would have major consequences for Alaska and much of the air travel between North America and Asia. Massive volcanic ash falls from eruptions can blanket vegetation in the summer, result in early snow melt in winter, clog streams causing flooding and/or fish kills and have numerous other short-term environmental effects. Depending on the type of ash, it can also contribute to the fertility of soils by adding phosphorus.

A number of active volcanoes are visible from Anchorage and the Kenai Peninsula. Recent eruptions include Mount St. Augustine in 1976 and 1986, Mount Redoubt in 1989, and Mount Spurr in 1953 and 1992. Each of these eruptions has deposited volcanic ash and residents of southcentral Alaska are well aware of the problems caused by residual ash. The small 1992 eruption of Mount Spurr dumped many tons of ash (2 to 3 pounds per square yard) on Anchorage, leaving a gritty layer 1/4 inch deep over the city. Years later, residents still contend with the effects of that eruption, as wind on dry days blows around the ash and causes respiratory distress for some individuals. Ash gets tracked into buildings and leaves a legacy of accelerated mechanical wear due to its abrasiveness.

On the Alaska Peninsula, major volcanic eruptions have literally reshaped the land. An excellent example is Aniakchak Crater. Created in a series of eruptions between about 4,500 and 3,500 years ago, ash-flows from this eruption traveled in many directions. Prior to the eruption, the Pacific Coast east of Aniakchak Crater was probably fiord-indented, like the modern coast to the north and south. Today at Amber and Aniakchak bays, eruption deposits fill these former fiords, creating broad, flat valleys drained by meandering rivers. Similar effects can be seen near Mount Veniaminov and near Roundtop on Unimak Island.

■ *Earthquakes*

In 1964, one of the strongest earthquakes to hit North America in historical times had its epicenter about 70 miles east of Anchorage in Prince William Sound. Extensive damage along the coast of Alaska from Kodiak to Cordova resulted. The community of Chenega was destroyed and Anchorage, Seward, Whittier and Valdez suffered major damage. Subsidence or uplift of large areas creates marine terraces, can change the flow of rivers and streams, and disturb or destroy plant and animal communities. Records from Russian America describe other major earthquakes such as one in Kodiak in 1792.

Earthquakes are a fact of life in southern Alaska and major earthquakes (greater than magnitude 7) can be expected every 50 to 100 years here. Fortunately, because large areas of Alaska are undeveloped, the cultural damage from most earthquakes is limited.

■ *Conclusion*

The nature of Alaska's construction serves to define the form of the terrain. Today, glaciers no longer cover large parts of Alaska, although remaining glaciers continue to sculpt areas of the Alaska-Aleutian Range and coastal mountains. Volcanoes continue to erupt and future large eruptions are likely. The slow, continuous tectonic reshaping of Alaska leaves its mark in generally imperceptible ways in the timescale of humans, although the occasional large earthquake reminds us of its effects. The land in Alaska is incredibly dynamic; past and ongoing changes have far-reaching effects on the plants and animals that inhabit Alaska. ■

These Precambrian rocks near Beaver Creek in the Interior are part of the original North American plate. Alaska consists primarily of this plate and of sections of land formed elsewhere that have been carried to Alaska by movement of plates in the Earth's crust. (Florence R. Weber)

The Terrible Lizards

By L.J. Campbell

Thousands of ancient fossils fill rows of metal cabinets in a cavernous storeroom beneath the University of Alaska Museum in Fairbanks. Some of the fossils, dark brown and shiny clean, sit in Styrofoam-lined boxes and little glass vials wearing neatly printed identifying labels. Many are bones from legs, shoulders, jaws, spines, skulls and tails. Some are smooth, thin rods of hardened tendons. A few are pointed cones edged with knife-sharp serrations, teeth from meat-eaters. Hundreds more fossils in these cabinets have not yet been cleaned, labeled or identified. They sit still bundled in layers of white tissue, protective wraps applied when they were dug out of riverbanks.

These fossils, only a sample of what waits to be unearthed in Alaska, have excited a worldwide community of scientists who study ancient Earth.

That's because these fossils come from dinosaurs, a diverse group of plant-eating and meat-eating vertebrates that lived during the Mesozoic. The Mesozoic lasted 245 million to 65 million years ago, long before humans appeared. It spanned three periods: the Triassic, Jurassic and Cretaceous. Alaska's dinosaurs found so far lived during the Cretaceous.

The word dinosaur means "terrible lizard," a name bestowed in 1841 by English anatomist Richard Owen to describe a growing assortment of then-new fossil finds, giant bones and teeth from seemingly related but previously unknown creatures. Owen and other scientists at the time imagined dinosaurs as giant reptiles. They assigned dinosaurs many characteristics of modern reptiles, including cold-blooded metabolisms. Unlike mammals, which are warm-blooded, reptiles do not maintain a constant body temperature. Instead, their body temperatures approximate that of their surroundings. They cool down rapidly and become sluggish in the cold; they are more active in the heat.

So dinosaurs were portrayed lumbering through the ancient tropics, living in warm climates like where reptiles are found today. This view prevailed largely unchallenged until the early 1970s, when paleontologist Robert Bakker championed a heresy first suggested by John Ostrom — dinosaurs were warm-blooded.

Dinosaurs dominated Earth as the reigning land creatures throughout the Mesozoic and to do this, Bakker said, dinosaurs had to have been more similar to warm-blooded mammals than to cold-blooded reptiles. The smaller meat-eating dinosaurs probably preyed on Mesozoic rodent-sized mammals, effectively suppressing development of larger mammalian species, Bakker argued. To have been constantly on the move, searching the land for plants and animals to eat, dinosaurs needed the higher energy levels of warm-bloodedness, he theorized.

Whatever their metabolism, dinosaurs died out at the end of the Mesozoic. They disappeared in a mass extinction that exterminated about 75 percent of life forms. Perhaps a giant meteorite collided with Earth, proposed

LEFT: *Crafty* Troodon *with its large eyes may have hunted at night, when small mammals were most active.* Troodon *is sometimes called the Cretaceous coyote. (Jan Sovak, reprinted from* The Last Great Dinosaurs, *courtesy Red Deer College Press)*

RIGHT: Pachyrhinosaurus *ranged from Alberta, Canada, where more than 2,200 bones have been found, to Alaska's North Slope where a single skull was found in 1988.* Pachyrhinosaurus *had a thick, bony growth on its face, although whether it supported a nose horn is open to interpretation. The bony nose plate on the Alaska* Pachyrhinosaurus *covers a greater area than that of similar dinosaurs found in Alberta, says paleontologist Philip Currie who has extensively studied Alberta's specimens. He suggests Alaska's critter may represent a new* Pachyrhinosaurus *species. This* Pachyrhinosaurus *is drawn from Alberta finds. (Jan Sovak, reprinted from* The Last Great Dinosaurs, *courtesy Red Deer College Press)*

University of California physicist Luis Alvarez in 1980. Clouds of dust billowed into the atmosphere on impact and the world plunged into a dark, cold, killing spell, he suggested. The evidence? Presence of iridium, an element relatively abundant in some types of meteorites from outer space but rare on Earth except in a thin, discontinuous layer of clay deposited about 65 million years ago. This clay has been found at a number of places around the world. A 121-mile-diameter crater since discovered on the Yucatan Peninsula in Mexico is thought to be the point of impact.

But back to Alaska's dinosaurs. In 1984, news hit the paleontological community that the North Slope of Alaska contained dinosaur bones. This was the first hard evidence that groups of dinosaurs had lived in polar regions. Dinosaur bones since have been found in Antarctica and Australia, also at polar latitudes during the Cretaceous.

At the time of the dinosaurs, Alaska's North Slope was probably several hundred miles farther north than today. Although the world was warmer during dinosaur times and the poles were ice-free, the high-latitude lands were still seasonally cold with periods of snow and several months of darkness. By the end of the Mesozoic, the climate in Alaska's northlands was similar to Anchorage's today with cool summers and freezing winters.

Dinosaurs not only inhabited places that today have climates inhospitable to modern reptiles, they experienced conditions similar to what impact theorists say caused their extinction.

"When large quantities of dinosaurs were found in Alaska, it was a mind bender," says Philip Currie, a renowned paleontologist at the Royal Tyrrell Museum in Alberta, Canada. "The implications? As fertile as your imagination... for dinosaur physiology, behavior and extinction."

How did dinosaurs survive high-latitude climates? Were they truly more similar in metabolism to mammals than reptiles? Did they migrate south to warmer winter weather? Did they travel an east-west dinosaur beltway? And what can Alaska's dinosaur bones contribute to discussions about early Earth?

Alaska's best known dinosaur beds on the

Colville River in northern Alaska hold tons of bones, along with leaf and pollen fossils that help scientists determine what the climate and vegetation were like during the late Cretaceous.

An added bonus is that many of the dinosaur remains are exceptionally well-preserved. In some cases, these bones have undergone little of the mineralization that makes fossil rock. They are still mostly bone. Alaska's dinosaur bones hold promise for scientists trying to isolate DNA to study genetics and define relationships between dinosaurs and modern-day reptiles and birds. The near original condition of some of Alaska's dinosaur bones, says Currie, triggered the idea of extracting dinosaur genetic material from fossils, an idea played out in the book and 1993 box office hit "Jurassic Park," about a theme park island populated with dinosaurs cloned from fossilized DNA.

The Liscomb bone bed on the Colville River contains piles of hadrosaur bones. Roland A. Gangloff, curator of the Earth Sciences collection at the University of Alaska Museum in Fairbanks, has led excavations of the site and suggests the following scenario depicted in these sketches: **LEFT:** *Adult and juvenile* Edmontosaurus *hadrosaurs are crossing a small waterway on the North Slope delta at flood stage, when the young get pulled away by currents and drown.* **LOWER LEFT:** *The meat-eating* Troodon *and* Albertosaurus *scavenge on the hadrosaur carcasses, which float in ponds left by the receding flood waters. Hadrosaur body parts fall to the muddy bottom.* **BELOW:** *A schematic stratigraphic section of the Liscomb bone bed in place today on the Colville River. (Tom Stewart illustrations; courtesy of Roland Gangloff, University of Alaska Museum)*

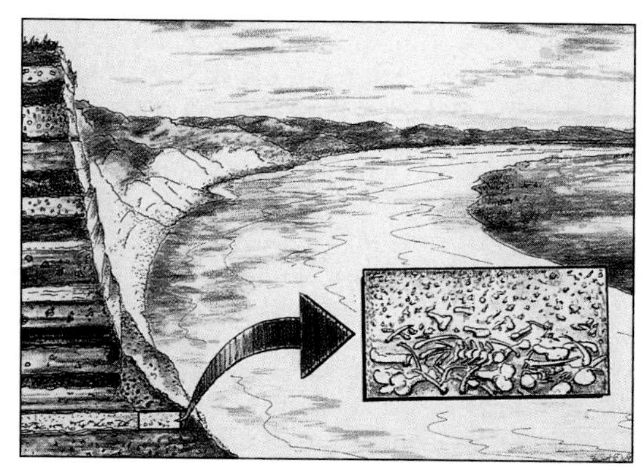

The search for dinosaurs in Alaska began in earnest in 1984. That summer a handful of geologists from the U.S. Geological Survey, led by Elizabeth Brouwers and David Carter, floated the Colville River on Alaska's North Slope. They were trying to pinpoint the location where dinosaur bones had been found more than 20 years earlier by Shell Oil Co. geologist Robert Liscomb.

Finding dinosaur bones is often more accidental than anything, and the Liscomb discovery is the most notable of several serendipitous finds in Alaska's dinosaur story.

In 1961, Liscomb had taken bones weathering out of a Colville River bank back to his office in California. Then he died the next year in a rock slide while working on an island in the Gulf of Alaska. His office was cleaned out, and his collection moved into the company's archives in Houston, Texas. The bones sat in oblivion until 1983, when Shell geologist Richard Emmons saw them while surveying fossils in storage. He sent them to C.A. Repenning of the USGS, who recognized them as dinosaur fossils and sent them to a leading dinosaur expert, Wann Langston, Jr., at the University of Texas.

Meanwhile, several other geologists had been trying to publicize their discoveries of dinosaur footprints and skin impressions in northern Alaska. In 1975, a geologist working on the Colville River had found a tree-toed track in the rocks, but an article about this dinosaur footprint was rejected as too speculative by a scientific journal. Then in 1978, geologists H.W. Roehler and G.D. Stricker had announced their find of dinosaur footprints and skin impressions on the Kokolik River. Their paper was not published until 1984, the year Liscomb's dinosaur collection was recognized, and another three years passed before an article about the three-toed footprint finally appeared.

Up to this time, the few other reports of dinosaur-era fossils in Alaska had been of sea

Alaska's Dinosaurs

Albertosaurus (Al-BER-toh-SORE-us): Large meat-eating, two-legged tyrannosaurid dinosaur; sharp teeth and claws; hunter/scavenger. Name means "Alberta lizard."
Size: 30 feet long; 10 feet tall; more than 3 tons. Alaska's tyrannosaurs appear to be smaller than this standard.
Evidence: Several teeth, a few neck, tail and foot bones, from Liscomb bed near Ocean Point on Colville River; teeth also found at confluence of Ninuluk and Colville rivers; finds may represent more than one species.
Age of finds: Late Cretaceous, 68 million to 72 million years old.

Edmontonia (ED-mon-TOE-nee-uh) Plant-eating, four-legged nodosaur, with narrow snout and thick, leathery armorlike plates across its back edged in spikes. Name means "from Edmonton."
Size: 23 feet long; 6 feet tall; 4 tons.
Evidence: Partial skull with two teeth found in western Talkeetna Mountains.
Age of finds: Late Cretaceous, 68 million to 72 million years old.

Edmontosaurus (Ed-MON-toh-SORE-us): Plant-eating, four-legged hadrosaurs; also known as duckbill dinosaur because of distinctive broad, flat nose; may have inflated loose skin over its nose to appear more threatening or attract a mate. Name means "Edmonton lizard."
Size: 40 feet or more long; 10 feet tall; 3 tons. Alaska's hadrosaurs possibly smaller than this standard adult size.
Evidence: Teeth, many bones, skin impressions, foot prints from various North Slope locations; the most abundant dinosaur bones in Liscomb bed on Colville River, also from beds downriver; hadrosaur bones represent about 95 percent of Alaska dinosaur collections; finds may include a crested-type of hadrosaur, yet to be determined.
Age of finds: Late Cretaceous, 68 million to 72 million years old.

Pachyrhinosaurus (Pack-ee-RINE-uh-sore-us): Plant-eating, four-legged ceratopsian dinosaur; thick hornless bone mass on nose led to belief it was the only ceratopsian without a nose horn; experts now consider a lightweight horn possible; neck frill edged with spikes. Name means "thick-nosed lizard."
Size: 18 feet long; 7 feet tall; 4 tons.
Evidence: Skull discovered upriver from main Liscomb bed on Colville River in 1988; possible horn core found nearby; other possible ceratopsian remains may be more of same or a larger relative, perhaps three-horned *Triceratops*.
Age of finds: Late Cretaceous, 68 million to 72 million years old.

Saurornitholestes (SORE-or-NITH-o-LESS-teez): Meat-eating, two-legged dromaeosaurid dinosaur; fast moving hunter/scavenger with slashing claw on back foot. Name means "bird robber."
Size: 6 feet long; 4 feet tall; 100 pounds.
Evidence: Tail vertebrate; teeth found along Colville River beach upriver from Liscomb bed.
Age of finds: Late Cretaceous, 68 million to 72 million years old.

Thescelosaurus (The-SKEL-o-SORE-us): Plant-eating, two-legged dinosaur of the agile hypsilophodontid (hip-suh-LOW-fuh-don-tid) family; a network of tendons supported its long tail straight behind for balance when running. Name means "pretty lizard."
Size: 11 feet long; less than 5 feet tall; 200 pounds.
Evidence: Teeth and toe bone found on Colville River, 5 miles upriver from Liscomb bed.
Age of finds: Mid to late Late Cretaceous, 78 million to 82 million years ago.

Troodon (TRU-oh-don): Meat-eating, two-legged dinosaur with large brain cavity; opposable digits on hands; large eyes; fast-moving hunter/scavenger with sicklelike claw on back foot; bones lightweight, filled with air pockets like bird bones today. Name means "wounding tooth."
Size: 8 feet long; 6 feet tall; several hundred pounds in weight.
Evidence: Numerous teeth, some bones from Liscomb bed; often found in association with hadrosaurs.
Age of finds: Late Cretaceous, 68 million to 72 million years old.

Adapted from summary by Bob King, Bureau of Land Management; information provided by Roland Gangloff, and The Last Great Dinosaurs *(1990).*

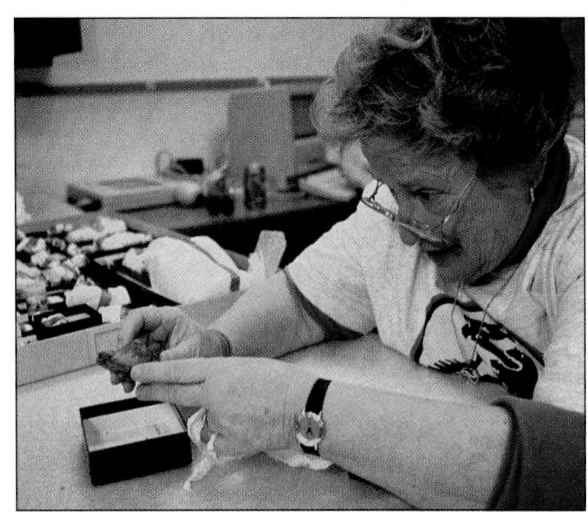

TOP LEFT: *Roland Gangloff, curator of the Earth Sciences collection, University of Alaska Museum, shows some of the North Slope dinosaur bones in the museum's collection. Gangloff and others have collected bones for the past seven years along the Colville River, mostly from a short section of the 200-foot-long Liscomb bed. He says Alaska's wealth of dinosaur relics are only starting to be discovered. (L.J. Campbell, staff)*

ABOVE: *Barb Gorman unwraps a dinosaur bone, one of many from the nearby tray still wearing a protective covering of tissue paper. These dinosaur fossils were found during the 1993 excavation of the Liscomb bone bed on the Colville River, near Ocean Point. (Steven Seiller)*

LOWER LEFT: *Paleontologist William Clemens, of the University of California Berkeley's Museum of Paleontology, works at the Liscomb bone bed during the 1988 field season. Dr. Clemens led several expeditions here subsequent to the bed's rediscovery in 1984 and today is actively involved in analyzing Alaska's dinosaur and Cretaceous-era small mammal fossils. (Mark Goodwin)*

dragons, or swimming reptiles, not the land-abiding dinosaurs. In 1922, USGS geologist W.R. Smith, who was working near the mouth of the Kejulik River, collected fossils of a plesiosaur, a sea reptile from the Jurassic period. In 1950, Irv Trailleur of the USGS discovered the skeleton of a dolphinlike ichthyosaur of Triassic vintage on a remote outcropping in the central Brooks Range foothills. He revisited the site in 1968 and two years later geologist Gil Mull, then with Exxon, flew in by helicopter and erected a sign identifying the fossil and asking visitors to leave it alone for future study. Also about 1969, USGS's Repenning identified a skeleton of a mixosaur, a type of ichthyosaur, found in limestone on Gravina Island in southeastern Alaska.

As it turned out, the 1984 float trip down the Colville launched the first concentrated effort to collect and study Alaska's dinosaur fossils. The USGS crews successfully located Liscomb's original bone bed on federal property near Ocean Point, a bend in the river about 25 miles from the coast. The bed was rich in dinosaur bones. The next year, with funding from the USGS, paleontologist William Clemens from University of California Berkeley, led a team, including University of Alaska paleontologist Carol Allison, to start excavating the Liscomb bone bed. They also prospected nearby outcroppings, where they found numerous other vertebrate fossils, teeth and skin imprints.

The excitement generated by these initial finds has yet to subside, and significant new discoveries have come from the Colville River's Ocean Point almost each year since. Although crews led by university researchers have excavated the Ocean Point quarries almost each summer since 1987, they have hardly scratched the surface.

"This place is unbelievably rich," says Roland Gangloff, curator of the Earth Sciences collection at the University of Alaska Museum in Fairbanks. He has led field work at Ocean

Point since 1991 and is Alaska's resident dinosaur expert.

"I'll stick my neck out. It may be one of the greatest Cretaceous vertebrate regions in the world," he asserted during a 1993 work session at the museum as volunteers carefully cleaned and catalogued some of the many trays of bones in storage. "When this is explored and documented, I believe we'll find hundreds of sites on the Colville alone. It could take us 30 years of work just to wash off the beach deposits, 100 years to properly assess what we have out there."

The thought of what other dinosaur sites might exist in Alaska is something the experts can only wonder about. "If Alaska was more accessible, there might be more field prospecting," says geologist Mull, now with the Alaska Geological Survey. "In a lot of these places, you need a helicopter and that's $500 to $600 a flight hour. It's awfully expensive."

Even returning to places of previous discoveries is difficult. Gangloff would like to investigate footprints of two meat-eating dinosaurs near Black Lake on the Alaska Peninsula that were photographed in 1975 by Phillips Oil Co. geologists. So far, however, he has had little luck determining the exact location and even less money to buy helicopter time to scope the area.

Worldwide, dinosaur research gets little funding, says Alberta's Currie. He estimates that as few as 30 paleontologists hold paid positions to study dinosaurs with a collective budget of less than $1 million annually. Field work in Alaska has been funded by several sources including the National Science Foundation, British Petroleum and the Eisenhower Mathematics and Science grant programs.

Excavations at the Ocean Point site were suspended in 1994, when the latest round of money ran out. Gangloff instead led a month-long boat trip down the Colville River in August 1994 to scout several promising new sites he'd found the previous summer, including a stretch of badlands several miles inland. One of the places he planned to visit was the location of another accidental find several years ago by an oil company geologist.

The geologist was sitting on a rock, waiting for a helicopter pickup north of Umiat, when he noticed an unusual shape jutting from the stone. With his hammer, he knocked loose a serrated tooth, which he took home and gave to his son. He later met Gangloff at a geological meeting in Anchorage and described the tooth. Gangloff thought it had probably come from a

Dinosaurs of Late Cretaceous Alaska *(overleaf). This illustration shows seven dinosaurs that lived in Alaska during the Cretaceous.* Thescelosaurus *(1), upper left, lived on the North Slope, as did* Saurornitholestes*(2),* Pachyrhinosaurus *(3),* Edmontosaurus *(4),* Troodon *(6) and* Albertosaurus *(5), shown on a river delta. The agile* Saurornitholestes *eats a sturgeonlike fish while* Troodon *chases after a loonlike* Hesperornis *(8), but eyes perhaps an easier catch, a rodentlike mammal (9). Teeth of several small mammals have been found associated with Alaska's North Slope dinosaurs. These mammal teeth resemble those of shrewlike* Cimolodon nitidus, *ratlike* Gypsonictops sp., *and a small marsupial. A partial skull of* Edmontonia *(7), lower right corner, was found in the Talkeetna Mountains, so this dinosaur is shown along a rugged river bank with vegetation similar to what might have been found in these volcanic mountains.*

Leaf and wood fossils found on the North Slope alongside Alaska's dinosaur bones give scientists a better idea of the Late Cretaceous climate and vegetation. Arctic Alaska was warmer then than today, although the region still experienced seasonal darkness and freezing temperatures. During the early Late Cretaceous, a temperate climate in the north allowed lush growth with a diversity of plants. An extinct, deciduous, broad-leaved conifer, Podozamites, *flourished in lower flood plains. Metasequoia-type trees and deciduous conifers, similar to pond cypresses, grew alongside viney cycads and shrubby gingkoes, with* Sphenopteris *ferns in the understory and* Equisetum *horsetails pioneering disturbed ground. Several millions of years later nearer the end of the Cretaceous, a time coinciding with most of Alaska's dinosaur fossils, the climate had cooled slightly and plant diversity had declined. North Slope forests consisted primarily of conifers similar to dawn redwoods, with ferns, weedy flowering plants and horsetails making up the understory, as depicted in the central part of this painting. (Painting by Tom Stewart)*

STEWART 94

meat-eating dinosaur, a rare find even in known bone beds. The tooth was later identified as a tyrannosaurid tooth, probably from a small cousin to the gargantuan *Tyrannosaurus Rex* of more southerly locales. Gangloff hoped to find more dinosaur relics at this and other sites during his 1994 prospecting trip.

While the North Slope sites have received the bulk of funding and research time, one other location has gotten some attention because it has been relatively easy to reach.

Gangloff and University of Alaska Anchorage professor Anne Pasch are collaborating on work in the Talkeetna Mountains following the 1990 discovery by a family of amateur fossil collectors of a nodosaurid ankylosaur skull.

But little is known about other places in Alaska where dinosaurs may have lived millions of years ago. "At this point, we just have bits and pieces," says Gangloff, "but there's a vast area to be investigated in Alaska."

So far, seven types of dinosaurs have been identified from bones collected in Alaska. They seem most similar to dinosaurs found in the extensive badlands of Dinosaur Provincial Park in Alberta, Canada, one of the richest bone sites in North America. Many suppositions about Alaska's dinosaurs stem from work during the past three decades by paleontologists with the Royal Tyrrell Museum in Drumheller. The museum is the depository for bones from the dinosaur park.

Four of Alaska's dinosaurs were plant-eating types and three were meat-eaters. Most of the identifications come from teeth, vertebrae, leg and toe bones. No complete dinosaur skull or

ABOVE: Edmontosaurus *hadrosaurs roamed Alaska's North Slope during the late Cretaceous. Most of the dinosaur bones found here have been of this type. The animal with the fish in its bill is a* Quetzalcoatlus, *the biggest flying animal yet known, with a wingspan up to 48 feet. Smaller* Pteranodons *fly in the background. No evidence of these flying species has been found in Alaska. (Jan Sovak, reprinted from* The Last Great Dinosaurs*; courtesy Red Deer College Press)*

RIGHT: Albertosaurus *preyed on other dinosaurs by hunting and scavenging. It was kin to the more southerly* Tyrannosaurus rex, *although it was smaller, longer limbed and probably faster. (Vladimir Krb; courtesy of Royal Tyrrell Museum)*

LEFT: Pachyrhinosaurus *probably gathered in herds and may have traveled many miles across the north searching for fresh feeding grounds or places to breed and raise their young. In his illustration of such an arctic migration, artist Vladimir Krb shows* Troodon *in the foreground and* Albertosaurus *at river's edge, both poised to prey on small or weak members of the herd. (Vladimir Krb)*

BELOW: *This* Edmontonia *nodosaur shows its clubless tail as it disappears into the forest. Spikes sticking out from its sides and bony plates covering its wide body provided armorlike protection. It may have dropped to the ground to protect its soft underbelly when attacked. Tiny teeth indicate that* Edmontonia *may have fed on soft water plants. (Jan Sovak, reprinted from* The Last Great Dinosaurs, *courtesy Red Deer College Press)*

skeleton has been found in Alaska, although two partial skulls from different types of dinosaurs have been recovered. The collection does include enough bones to reconstruct a nearly complete composite skull and skeleton of the duckbill hadrosaur *Edmontosaurus*, the most common dinosaur found on the North Slope.

Hadrosaurs were a family of large four-legged, plant-eating dinosaurs that lived worldwide during the late Cretaceous. Hadrosaurs are often called duckbills because of their broad, flat noses. Unlike ducks, they had hundreds of teeth that formed a rasplike grinding surface, adapted to eating tough plants. They could stand on their back legs to reach overhead foliage, although they probably traveled on all four legs in a rocking gate, since their front legs were shorter. They had hoofed toes. Many hadrosaurs had head ornamentations or crests, hollow structures thought to resonate sounds. The genus *Edmontosaurus* found in Alaska had no crest, but may have amplified its calls using the high palate of its mouth. *Edmontosaurus* was the largest of the hadrosaurs and may have weighed several tons. Its large eye sockets suggest it had eyes perhaps 4 inches wide, which would have given it good eyesight.

Edmontosaurus makes up about 95 percent of the identified bones from the North Slope. Hadrosaurs were social animals that gathered in herds. Many *Edmontosaurus* bones have been found in piles, as if groups of them had died in a flash flood on a river delta near the ocean, says Gangloff.

Many of Alaska's *Edmontosaurus* bones are much smaller than those found in more southerly locales. Alaska's hadrosaurs may have been a diminutive version, their smaller size some sort of adaptation to the polar environment. Or the bones may represent only juvenile hadrosaurs that were more susceptible than adults to whatever killed them.

Alberta's Currie suggests Alaska's hadrosaurs

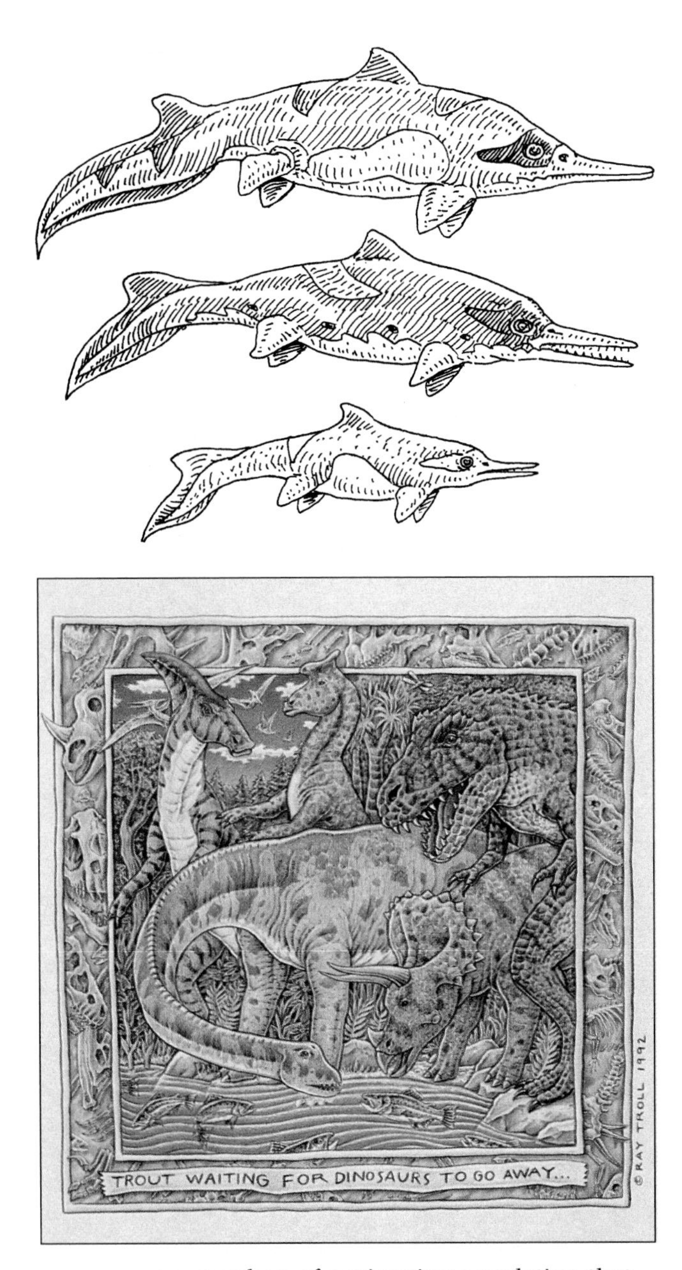

LEFT: *The only Mesozoic vertebrate found so far in Southeast Alaska is this* Mixosaurus, *a type of ichthyosaur swimming reptile from the Triassic period. This dolphinlike creature was 3 to 4 feet long. Its snout was lined with sharp teeth. A section of vertebrate and ribs were embedded in limestone on Gravina Island and identified in 1969. (Illustration by Ray Troll)*

LOWER LEFT: *Despite their disappearance from the land some 65 million years ago, dinosaurs live on — in movies and toy stores, on coffee mugs and T-shirts. Ketchikan, Alaska, artist Ray Troll steps back in prehistoric time with this T-shirt tribute: "Trout waiting for dinosaurs to go away." (Ray Troll)*

were members of a migrating population that spent winters in warmer southerly climates. At the same time, UAF's Gangloff is excited about discoveries on the North Slope of a few tiny hadrosaur finger bones, perhaps from hatchling hadrosaurs. While no nests or eggs have been discovered in Alaska, Gangloff thinks these bones suggest that hadrosaurs might have been year-round residents or came to the North Slope to have their young, as caribou do today.

Found among the hadrosaur bones have been bones and teeth of several other types of dinosaurs. These include at least one type each of tyrannosaurid, dromaeosaurid and troodontid.

Tyrannosaurids were a family of large meat-eating dinosaurs that lived during the late Cretaceous. They had huge heads and mouths full of sharp, serrated teeth. They stood upright on powerful back legs. Their short front legs were useless for locomotion and probably used only to hold prey. The *Tyrannosaurus rex,* or tyrant lizard, was the largest in the family. The Alaska specimens are the genus *Albertosaurus*, a smaller version of the *T-rex*.

The dromaeosaurid dinosaurs were small, active dinosaurs, also upright in stance. Their back feet had an oversized claw, possibly used for slashing attacks on other animals. The Alaska specimen has been identified as a dromaeosaurid of the genus *Saurornitholestes*.

Troodon was a small, fleet meat-eater that also stood upright on its back feet. *Troodon* is considered the smartest dinosaur based on its large brain cavity.

The least known of the North Slope dinosaurs is a type of small plant-eating hypsilophodontid, tentatively identified from teeth and toe bones as *Thescelosaurus*. Hypsilophodontids first appeared in the late Jurassic and lived through the Cretaceous. *Thescelosaurus* was a late Cretaceous version. They were one of a much larger group collectively known as Ornithopoda, which also included the hadrosaurs.

Berkeley's Clemens and graduate student Gayle Nelms think these small creatures might have been permanent residents of the North Slope, because some of the teeth were from young hypsilophodontids too small to have traveled thousands of miles. Nelms in 1994 was completing her doctoral dissertation about the North Slope finds.

One of the remarkable finds from the North Slope came in 1988 when University of California researcher Howard Hutchison took a break from the Ocean Point dig. He sauntered about a mile up the beach where he saw a protruding skull. It was about 75 percent intact and identified as *Pachyrhinosaurus*, a type of ceratopsian and a rare relative of the better known, three-horned *Triceratops*. Ceratopsians included all the various horned dinosaurs that lived in the late Cretaceous. They were four-legged with bulky bodies similar to modern rhinoceroses. Most of them had horns on their noses and above their eyes, and all of them had bony frills extending from their heads over their necks. Their mouths ended in a curved bony beak used to snip tough plants and branches.

All the known *Pachyrhinosaurus* fossils show a bony base, or "nose boss" and until recently, this dinosaur was thought to have been the only hornless ceratopsian. However, paleontologist Currie now speculates that

Pachyrhinosaurus had a nose horn as well as horns on its neck frill. Hutchison's find of a *Pachyrhinosaurus* in Alaska extended the creature's range, since it previously had been found only in Alberta.

The Ocean Point beds have yielded other dinosaur-era fossils including small marsupials and rodentlike mammals; a loonlike bird called *Hesperornis*; a turtle; a sturgeonlike fish; and a variety of advanced bony fish. Some of the unidentified dinosaur bones may belong to still other species of ceratopsians, tyrannosaurs and hadrosaurs.

Curiously, fossils of crocodilians and amphibians are conspicuously absent from Alaska's bone beds. These fossils are commonly found in warmer locations of comparable age. The Berkeley researchers think this indicates that dinosaurs were indeed adapted for the cold.

The single dinosaur from the Talkeetnas came to science on a path similar to that of Liscomb's North Slope find. It was discovered, then rediscovered about 20 years later.

John A. Luster, an old-time trapper and horseback hunting guide now in Wasilla, was checking his trap line in the western Talkeetnas one winter day back in 1968 when he found a big skull sticking out of a riverbank. When he got home, he told his family about the dinosaur he'd found. He offered a horn from his pocket as proof. But Luster liked tall tales and told them often to his 12 children. Inventing animals was his forte. Hanging from the top of a shed was a cow's skull, pigs teeth and assorted other parts that he called a "helifornis."

"When he started talking about dinosaurs, we figured it was just another helifornis," recalls his son John E. "We didn't believe him."

Two decades later, when Luster turned 90, he was still talking about the dinosaur, and still urging his family to go find it. Two of his sons and a daughter decided to humor him. They loaded their horses and headed behind him into the hills. "We found it just where he said it was," says John E. "It had eroded out of the bank and was just laying there."

They wrapped it in spare shirts and underwear and carried it out in an ice chest where it set a few days by the front door of the house while they decided what to do with it. The local newspaper ran a picture of them with the skull, and soon it was making the rounds of the university dinosaur experts.

Gangloff worked hours by hand to clean the skull in his laboratory at UAF. Using a microscope to see, he painstakingly removed the hard rock matrix that encrusted the skull and its crevices. He was rewarded by finding two teeth — one in almost perfect condition — lying in the palate where they had dropped out of the jaw. From those and the shape of the jaw, he tentatively identified it as nodosaurid ankylosaur, a type of four-legged, plant-eating armored dinosaur.

In 1922, W.R. Smith recovered plesiosaur fossils at the mouth of Kejulik River, Alaska Peninsula. In 1994, scientists determined those fossils, now in a museum, were from a short-necked, huge-headed pliosaur, a type of plesiosaur from the late Jurassic, related to the Elasmosaurus *shown here. (Vladimir Krb; courtesy of Royal Tyrrell Museum)*

These dinosaurs are similar in shape to modern armadillos, although much larger with various arrangements of bony plates and studs covering their backs and heads. The ankylosaurs bore bony knobs on their tails, probably wielded in defense like clubs; the nodosaurs lacked the tail knobs and may have hunkered down in place when threatened.

Gangloff sent the nodosaur skull to the

University of California Museum in Berkeley where Gayle Nelms, Mark Goodwin and Howard Hutchison confirmed the identification. Three casts of the skull were made for the university museums in Berkeley and Fairbanks and the Royal Tyrrell Museum in Alberta. The Alaska specimen is now classified as the genus *Edmontonia*. The original skull was eventually shipped back to Wasilla, where it now resides at the Dorothy Page Museum. Every now and then, its discoverer John Luster stops in to say hello.

Along with dinosaurs, the North Slope bone beds reveal a much different Alaska during the late Cretaceous.

ABOVE LEFT: *Teeth and a toe bone found on the Colville River about five miles from Ocean Point probably belonged to the plant-eating* Thescelosaurus *. This scene shows* Thescelosaurus *fleeing from a swamp fire, which would scar the land and allow new plant growth.* Thescelosaurus *legs were longer in the thigh than calf, unlike other members of the hypsilophondont family. (Jan Sovak, reprinted from* The Last Great Dinosaurs, *courtesy Red Deer College Press)*

ABOVE: Saurornitholestes *could have been a formidable hunter, fast, agile and equipped with sicklelike claws on its back feet. Scientists speculate that these dinosaurs may have hunted cooperatively in packs to bring down larger prey, based on finds in Alberta, Canada. (Jan Sovak, reprinted from* The Last Great Dinosaurs, *courtesy Red Deer College Press)*

Plant paleoecologists Robert Spicer, of Oxford University in England, and paleoclimatologist Judith Totman Parrish, of the University of Arizona in Tucson, have floated the Colville River several times, collecting plant fossils as part of a global look at prehistoric ecosystems and climate trends. Their work interprets a time span covering millions of years with emphasis on two distinct periods of the Cretaceous — the early Late Cretaceous and the late Late Cretaceous, covering broadly the last third of the Cretaceous period.

Their reconstructions, from wood, leaf and pollen fossils, show the North Slope as a lushly vegetated river delta with deciduous forests, not the boggy, tree-scarce plain it is today. Dinosaurs could have found plenty to eat during Alaska's polar summers.

More than 100 kinds of plants grew on the North Slope during the early Late Cretaceous, Parrish said. The climate then was like that of western Oregon today. There were thick forests of many types of cone-bearing conifers, trees that grew up to 27 inches in diameter. Fossilized leaves include that of Metasequoia, a conifer similar to modern-day dawn redwoods of China. The Alaska Cretaceous conifers apparently dropped their leaves in winter, a metabolic adaptation to survive the dark. The forest understory included shrublike gingkoes, and sycamorelike broadleaf trees grew along rivers and streams. Big peat swamps formed between the rivers channels. One of the hardiest plants was related to *Equisetum* and grew almost everywhere, particularly in wet, disturbed areas. *Equisetum*, or horsetails, is a common weed today.

During the middle Late Cretaceous, the sea rose and covered much of the North Slope. When the water receded, plants repopulated the region. But the climate had cooled several degrees, and the plant diversity had plummeted. By the end of the Cretaceous, fewer than 10 different plants are known from leaf fossils, although pollen fossils show an increase in weedy herbaceous plants. These small, non-woody plants probably didn't have leaves tough enough to survive fossilization, Parrish said. At the end of the Cretaceous, trees were smaller, averaging less than 20 inches in diameter. Growth rings in the fossil wood had a larger percentage of dark wood, which indicates a cooler growing season had moved in, she said.

How did northern dinosaurs survive the dark, cold winters? Perhaps they gorged on food in the fall and hibernated like some modern mammals. Perhaps they had layers of insulating fat or feathery coverings and subsisted on roots, or hunted small mammals. Perhaps some of them migrated to warmer climates with more plentiful food sources.

Currie, the Alberta paleontologist, suggests that the larger North Slope dinosaurs, like the hadrosaurs, may have migrated as caribou do today, traveling between the Arctic and more southerly feeding grounds. Clemens and Nelms, writing in the June 1993 issue of *Geology*, find long north-south migrations unlikely, particularly for the smaller, more solitary dinosaurs like the troodons and hypsilophodontid. Currie agrees that those dinosaurs may have been year-round residents.

Jumbles of bones found in the Liscomb beds suggest that hadrosaurs and ceratopsians periodically gathered in herds. They probably did have to move around to find enough food. A new dinosaur find in 1994 in Siberia, in ground similar to the North Slope bone beds, cements an earlier theory among some experts that the North Slope was part of an east-west corridor for dinosaurs traveling between North America and Asia over the land bridge that existed during the Late Cretaceous. This route would not have removed them from the winter dark zone, however.

This brings the experts back to the extinction theories. Perhaps the disappearance of Earth's dinosaurs was a response to more gradual changes. Perhaps dinosaurs from Asia crossed the land bridge during the Late Cretaceous and introduced fatal diseases to North American species, and the reverse also happened. Perhaps mammals contributed to the extinction by devouring dinosaur eggs. Perhaps dinosaurs failed to adapt to changing plants and climates.

"I think dinosaurs can tell us a lot," says Currie. "Kids want to know how big they got and how they lived. Adults want to know how they died. It's part of the morbid fascination with our own potential for extinction. We are causing tremendous reductions in diversity (of plants and animals). Are we setting the course for our own destruction? Until we solve that, dinosaurs will always be of interest."

Alaska's dinosaurs are contributing to these discussions, as well as to continually changing ideas about these creatures once simply dismissed as terrible lizards.

With each new fossil find, paleontologists tweak physical reconstructions of the dinosaurs, changing how they carried their tails, the length of their necks, their assorted horns, frills and crests. They imagine skin colors from subdued to bright, they guess how dinosaurs sounded, they even illustrate their eyes gleaming yellow. At the same time, they're offering greater variations on how these prehistoric animals behaved, crediting them with mammalianlike social structures that included herding, cooperative hunting and maternal instincts.

Alaska's dinosaurs and their counterparts from other polar regions are helping write a new chapter in the definition of what these creatures — and their world — were like. ■

BELOW: *In 1975, geologist Jim Fox, working along the Colville River, found this dinosaur track, among the first recognized evidence of dinosaurs in northern Alaska. At the time of its discovery, the find attracted little interest, in part because the bones Robert Liscomb had previously found in the bone bed that now bears his name had yet to be recognized as dinosaur material. (Gil Mull)*

RIGHT: *This photo shows 14 footprints from two different meat-eating dinosaurs left in an ancient streambed now turned to rock. The arrow points to a piece of petrified log. The exposure is about 12 feet high. This trackway was found in 1975 by Phillips Oil Co. geologists near Black Lake on the Alaska Peninsula, property now belonging to the Bristol Bay Native Corp. (Courtesy of Bristol Bay Native Corp.)*

Alaska Vegetation:

What the Fossil Record of the Past 20 Million Years Shows

By Dr. Thomas Ager

Editor's note: *Dr. Ager is a geologist in the Branch of Paleontology and Stratigraphy of the U.S. Geological Survey in Denver. He is currently leading a research team from the United States, Canada, Russia and the People's Republic of China that is gathering evidence about the history of high-latitude climate changes of northeast Asia, Alaska and northern Canada during the past about 20 million years.*

This paper draws on published information by Jack Wolfe, John Matthews, Jr., R. Dale Guthrie, John Barron, Thomas Hamilton, David Hopkins, Estella Leopold, James White, Mick Kunk and many others. Dr. Ager gratefully acknowledges their valuable contributions to our understanding of Alaska's geologic record.

■ *Alaska's Present-Day Vegetation*

Alaska encompasses 586,400 square miles spanning 20 degrees of latitude and 58 degrees of longitude. Topography is highly varied, ranging from broad coastal lowlands in the west and north, interior basins, and uplands and mountain ranges thousands of feet high. This broad range of latitudes and topographic features creates a wide variability of local and regional climates.

The long coastline also significantly influences regional climates. Cool, moist air from the North Pacific affects the climates of the southern and southeastern coasts, regions characterized by cool, wet summers and stormy, wet winters with relatively mild temperatures. Differences between winter and summer mean air temperatures are relatively small. In northern and western coastal Alaska, the maritime influence of the Arctic Ocean and the northern Bering Sea is less dramatic. Much colder seawater temperatures and the presence of sea ice during much of the year limit the warming influence of the adjacent seas, and less moisture is evaporated from a sea surface largely covered by ice.

In interior Alaska, the influence of oceanic air is minimal because of intervening mountain ranges to the north and to the south. Thus Alaska's interior has a continental climate characterized by low precipitation, warm summer temperatures, and cold winter temperatures, often falling below minus 50 degrees.

Regional and local variations in modern vegetation reflect several influences, such as latitude, altitude, topography, soil types, and local and regional climates. In contrast to the other 49 states, most of Alaska's vegetation still looks much as it did when the first explorers arrived. This allows us to interpret relationships between regional vegetation and climates without having to factor in large-scale human disturbances of vegetation.

If we travel across Alaska from south to north, we can see how dramatically the natural vegetation changes, and how the patterns reflect topographic and climatic influences. Dense coastal forests of Sitka spruce, western hemlock, mountain hemlock, red cedar and Alaska yellow cedar grow in coastal areas of

Southcentral and Southeast. The coastal vegetation also includes alders, devil's club, skunk cabbages, ferns and spongy carpets of moss. Tree trunks and branches in this rainy environment are festooned with mosses and lichens.

As we travel inland and ascend the mountains that parallel the coast, the forests gradually give way to thickets of alders and willows, and as we rise still higher, the shrub thickets are replaced by alpine tundra. With increasing altitude, alpine tundra plants become fewer and smaller, until little remains but lichen-encrusted or barren rock, snowfields and glaciers.

Alpine tundra vegetation is composed of patches of grasses and sedges, mosses, lichens and fragile-looking alpine flowers, many of which form low cushions. Few shrubs occur in the alpine tundra, and those that do tend to be

ABOVE: *Miniature willows grow in alpine tundra vegetation in the mountains of southern Alaska. (Thomas Ager)*

RIGHT: *Coastal forest vegetation, shown here near Juneau, includes a number of tree types such as Sitka spruce, western hemlock, red cedar and Alaska yellow cedar. Cottonwoods occur along some river bottoms, and shrubs include alders and devil's club. (Thomas Ager)*

BELOW RIGHT: *Coastal bluffs near Ninilchik yielded this fossil leaf of an alder shrub of late Miocene age. Plant fossils of this age from Cook Inlet areas are mostly of alders, poplars and willows, although conifer needles are also present. Global cooling in late Miocene time eliminated many types of trees and shrubs from Alaska that had been common during warmer periods. (Thomas Ager)*

small and grow low to the ground. Alpine tundra shrubs include several species of miniature willow, and low-growing plants such as crowberry, bearberry and alpine cranberry. Plants of the tundra can endure short growing seasons, cool summer temperatures and summer frosts, desiccating winds and frost-heaved, rocky soils.

Lowland tundra vegetation in northern Alaska is typically made up of tussock-forming sedges, grasses, low shrub willows, dwarf birch, small tundra flowers, aquatic plants and mosses and lichens. This photo shows Carter Creek, which flows from the Sadlerochit Mountains across the arctic coastal plain to Camden Bay. (Thomas Ager)

As we descend from the mountains into the interior, we encounter a much different landscape: broad valleys forested by white spruce and black spruce, with groves of aspens and paper birch trees on well-drained slopes, and balsam poplars, alders and willows in wetter valley bottoms. This spruce-dominated vegetation is called northern boreal forest or taiga. These trees grow far more slowly than those of coastal regions, and attain smaller sizes. Forests of the interior are adapted to warm, dry summers, extremely cold winters and relatively frequent fires. Many plants of the boreal forest can tolerate the cold, wet soils associated with permafrost—perennially frozen soils that occur in regions where the mean annual temperature falls below 32 degrees.

As we continue our journey northward, we cross the altitudinal and latitudinal transition from the boreal forests of the interior to the shrub vegetation, then alpine tundra of the Brooks Range, and finally to the arctic lowland tundra of the North Slope. Arctic tundra is composed largely of tussock-forming sedges, grasses, low willows, dwarf birches and small tundra flowering plants, mosses and lichens, and a variety of aquatic and semi-aquatic plants. Permafrost is close to the surface, resulting in cold, wet soils.

Alternatively, if we travel west from interior Alaska, we cross another transition from boreal forest to coastal, mostly lowland tundra. This vegetation boundary is not a result of increasing latitude or altitude as in northern Alaska, but reflects the cooling influence of the Bering and Chukchi seas. Summer temperatures are too cool for the survival of trees along most of Alaska's western coast. But summer temperatures are warmer than those in northern Alaska, so the lowland tundra of western Alaska is richer in shrub, herb and aquatic plant species.

How long have these regional patterns of vegetation types existed in Alaska? How did these patterns develop? And *if* Alaska vegetation differed in the past, why was it different and what did it look like? How has Alaska's vegetation reacted to past climate changes? These are some of the questions scientists have been seeking to answer. These are important questions because we need to understand the subtle relationships between organisms and their environments, particularly in sensitive high-latitude environments. We need to understand and predict the consequences of climate change and disturbance of ecosystems, whether caused by natural events or by an ever-expanding range of human activities. We need to put this information to use when government agencies make decisions that may have far-reaching and sometimes unforeseen consequences for our environment.

This chapter will present a summary of what is currently known about Alaska's vegetation and climate history during the past 20 million years. It is based on the study of plant fossils preserved in Alaska's geologic record—a priceless natural archive of information about the past.

■ *Understanding Relationships Between Modern Plants and Their Environment: Clues to Understanding Past Vegetation and Climates*

The genetics of each plant species largely control that plant's capacity to tolerate a particular range of environmental conditions, including soil type, available moisture, winter temperatures, summer temperatures, length of growing season, snow depth, shade, crowding from other plants, damage from insects and grazing animals and other factors.

A plant species' distribution reflects that plant's ability to occupy areas within its range of tolerance, but distribution also depends on the plant's ability to spread its seeds or spread by vegetative means. A plant's ability to compete for space, water, sunlight and nutrients also influences its potential abundance. The regional history of geology and climate also has an influence. Why is history important? A plant species cannot colonize or invade an area if it isn't present somewhere in the region to start with, or if it can't get there from somewhere else because of topographic or climatic barriers, or because its seed dispersal mechanisms are unfavorable for rapid dissemination.

Decades ago, ecologists realized that when environmental conditions change, individual plant species respond to those changes. Entire ecosystems do not respond as a unit, but as a sum of individual species' responses. Thus a given vegetation type, such as the boreal forest, is composed of many species with overlapping tolerances for a variety of environmental influences.

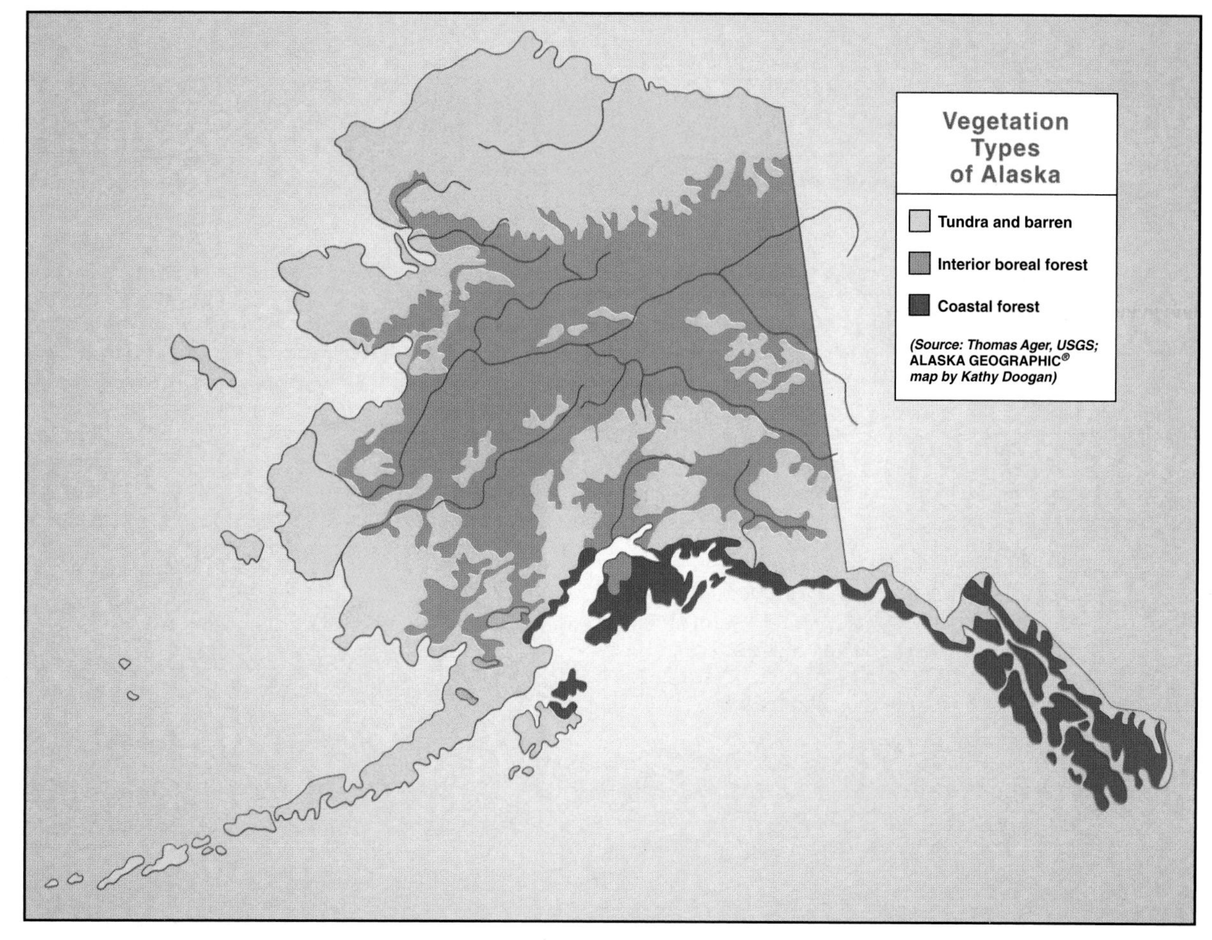

To illustrate, let us consider the plant species that are found today along the coasts of southern and southeastern Alaska. Each of those plant species has different tolerances for specific environmental factors, but the plants generally share an ability to tolerate cool, wet summers, and cool, wet and stormy winters. Some of the coastal taxa can tolerate salt-laden moisture that winds carry into the forests from the surf zone and sea.

Most species of trees and shrubs found in southern and southeastern Alaska coastal regions cannot tolerate the extreme winter temperatures and the dry summers found in the interior; thus their geographic spread is limited by the particular environmental conditions they require.

Conversely, many of the small plants found in a tundra environment are rarely encountered in lowland boreal forests, and many boreal forest taxa are unable to survive in the high mountains or in wet, coastal habitats.

These examples indicate that there is a general relationship between the kinds of vegetation one finds in a given area and the climatic and other environmental conditions that prevail within a region. There are

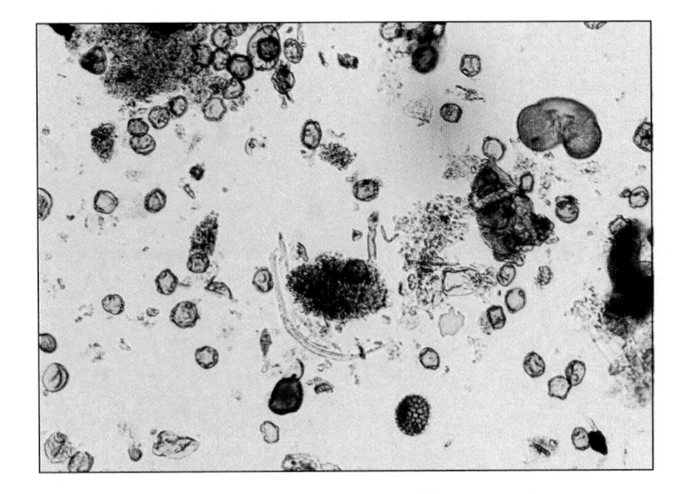

ABOVE: *This photo, taken with a camera-equipped microscope, shows some of the pollen types commonly found in lake sediments less than 8,000 years old in interior Alaska. Pollen types include spruce, alder, birch, grass, sedge and a type of blueberry. (Thomas Ager)*

RIGHT: *Lake sediment coring can be done in summer from floating platforms, such as shown here at Tungak Lake, a volcanic crater lake in the central Yukon delta. (William Dupre, courtesy of Thomas Ager)*

exceptions to this generalization, but it is a useful concept.

In spite of the exceptions, this general recognition of broad linkages between certain vegetation types and associated regional climatic conditions permits us to use fossil plant evidence to make inferences about past vegetation and climates. Plant fossils can therefore be used to reconstruct the broad characteristics of past vegetation, and the type of vegetation indicates past climate. This is one approach paleontologists have used to develop a history of Alaska vegetation and climate spanning millions of years.

In this chapter we focus on only the past 20 million years because the ecosystems that have existed during that time interval share enough similarities with modern ecosystems to be more easily interpretable than older floras. What is the nature of our evidence for reconstructing past environments? In most terrestrial environments today, plants are far more abundant, in terms of sheer numbers and biomass, than animals. This has been true for several hundred million years or more. Therefore there is a high probability that some plants will be preserved as fossils under a broad range of environmental conditions, and the fossil record reflects that. Plants are sometimes preserved as fossils in their entirety, but it is far more common to find fossil remains of only parts of a plant: leaves, seeds, stems or sometimes stumps or entire logs.

The most widespread plant remains that are often preserved are pollen and spores, microscopic-sized reproductive cells of plants. Conifers and many temperate broadleaf trees and shrubs, as well as many herbaceous plants, disperse their pollen by wind, an inefficient method, and therefore produce vast quantities of pollen.

Some flowering plants depend on insect pollinators, so they tend to produce less pollen and have fewer chances to be represented in the fossil record. The pollen and spores found preserved in the geologic record imperfectly represent the composition of past vegetation, but they provide important information about past environments.

The outer layers of pollen and spores are composed of an unusually durable substance, called sporopollenin, which gives these tiny plant remains the ability to endure as fossils if they are protected from oxidation by rapid burial. Thus pollen and spores deposited and quickly buried in lakes, peat bogs, swamps, or muddy sediments in rivers or estuaries are likely to be preserved as an essentially permanent archive about the flora and vegetation of the region. A flora is a list of plant types growing in a given area in the past or present; vegetation is the community of plants growing together in an area in the past or present.

If all pollen and spores looked alike, their ability to be preserved as fossils would be of little use for reconstructing past floras and vegetation. But there is an almost infinite variety of morphological variation—shapes, structure and ornamentation—in pollen and spores, and the morphology is often distinctive for a given plant type. Thus pine pollen looks dramatically different from grass pollen, and water lily pollen looks different from the spores of sphagnum moss. This variability gives us a powerful tool for reconstructing past floras and vegetation

I often think of my microscope as a time machine that permits me to capture glimpses of long-lost worlds. By scanning across a microscope slide strewn with fossil pollen grains and spores, scientists can identify pollen types and compile a list of plant types that existed near the site of deposition at a

The boreal forest vegetation of interior Alaska is typically composed of white spruce, black spruce, paper birch, aspen, balsam poplar, alder and willow with a ground cover of grasses, sedges, mosses, dwarf birch shrubs and low-growing blueberry bushes. This scene looks south across the Tanana Valley toward the Alaska Range. (Thomas Ager)

Some Characteristic Plant Types Found in the Fossil Record of Alaska for the Past 20 Million Years

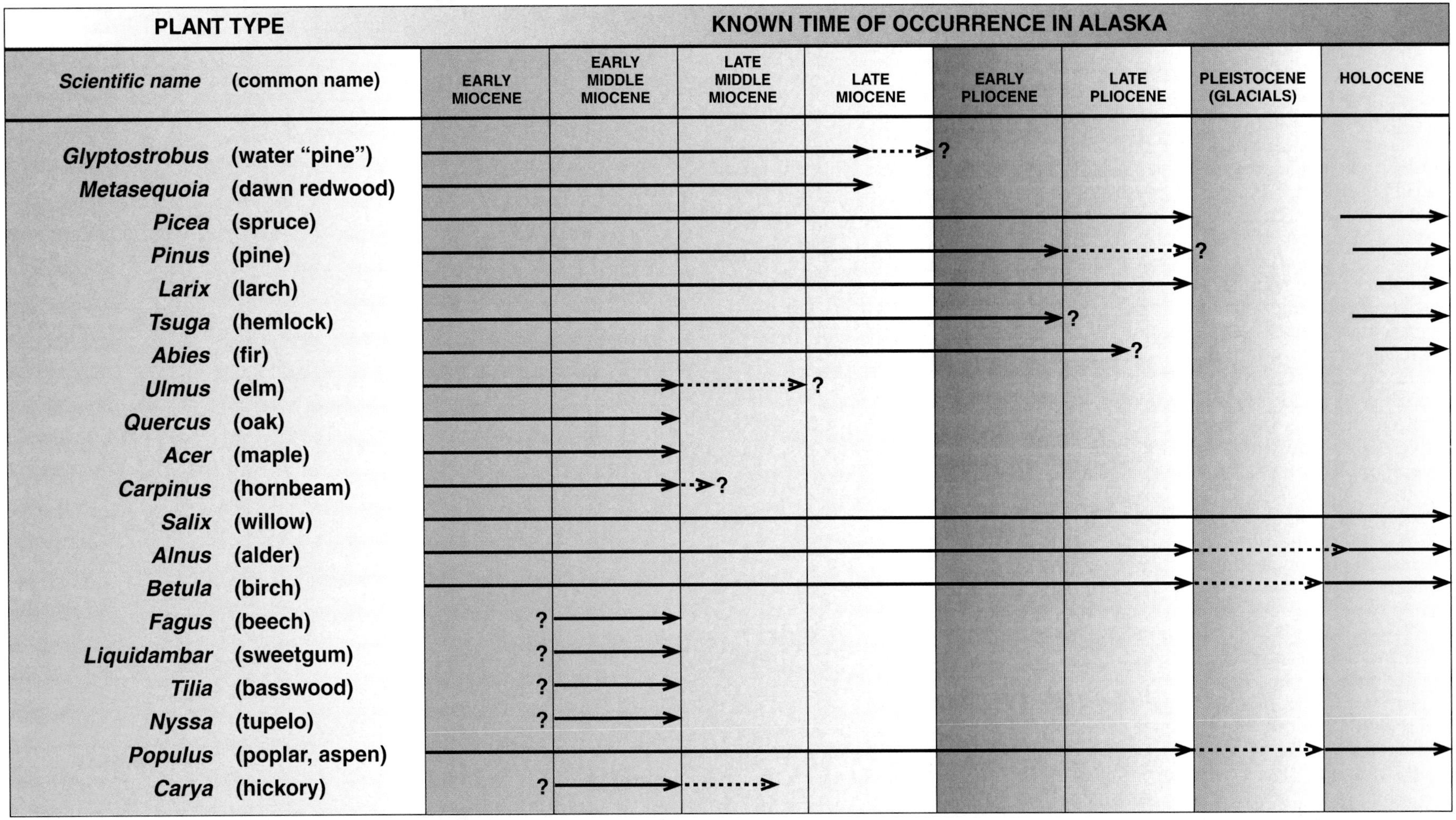

PLANT TYPE		KNOWN TIME OF OCCURRENCE IN ALASKA							
Scientific name	(common name)	EARLY MIOCENE	EARLY MIDDLE MIOCENE	LATE MIDDLE MIOCENE	LATE MIOCENE	EARLY PLIOCENE	LATE PLIOCENE	PLEISTOCENE (GLACIALS)	HOLOCENE
Glyptostrobus	(water "pine")								
Metasequoia	(dawn redwood)								
Picea	(spruce)								
Pinus	(pine)								
Larix	(larch)								
Tsuga	(hemlock)								
Abies	(fir)								
Ulmus	(elm)								
Quercus	(oak)								
Acer	(maple)								
Carpinus	(hornbeam)								
Salix	(willow)								
Alnus	(alder)								
Betula	(birch)								
Fagus	(beech)								
Liquidambar	(sweetgum)								
Tilia	(basswood)								
Nyssa	(tupelo)								
Populus	(poplar, aspen)								
Carya	(hickory)								

*This chart shows approximately when these plants grew in Alaska. Dashed lines show intervals of sporadic or rare occurrences. Question marks indicate uncertainty about first or last appearances. Herbaceous plants, not listed in this chart, were relatively rare in Alaska until the past 5 million years. Lowland vegetation types containing a high percentage of herbaceous plants, such as various tundra types, apparently developed in Alaska only during the past 2 million years or so. (*Source: *Thomas Ager, USGS; graphic by Kathy Doogan)*

particular time. Thus scientists can form a mental image of that local flora.

If scientists take the time to do a more detailed, quantitative analysis, it is possible to recover even more clues about the relative abundance of individual plant types in that past landscape. With that information, inferences can be drawn about the probable type of vegetation and climate that existed at that time.

Percentages of fossil pollen types in a sample from a past time interval can be compared with percentages of pollen from modern surface samples taken from sites with known vegetation types. Sometimes a close match can be found between the ancient and modern pollen assemblages, suggesting a close similarity between the fossil and modern vegetation types. However, often there are no close matches with modern vegetation types. Lack of a close match can make interpretations of past vegetation more challenging, but this lack also shows how ecosystems have adapted in the past to a wide range of environmental changes and chance events that may differ from conditions today.

Analysis of an individual sample of fossil pollen and spores can provide a snapshot of the vegetation of an area at one instant in the past. Analysis of a series of samples arranged in an age sequence, from older to younger, can tell us how floras and vegetation changed throughout time within a region.

The opportunity to visualize past environments through the study of fossil assemblages may be the closest thing we will ever have to experiencing time travel, at least outside of science fiction. It is an imperfect time machine, alas, because we can't see the entire landscape, nor can we see all the organisms that once dwelled there.

Let us board our time machine to journey back through time and space, back to the time before the ice ages, and before the time when the Alaska Range and the coastal mountains were formed. We will make several stops on our journey to see how Alaska's vegetation appeared at several key moments in the past 20 million years.

■ *Early Miocene*

Our first stop on this time-travel adventure will take us to a time in the early Miocene, about 20 million years ago. The location is near the town of Healy. Today this area has rugged mountains and forested valleys with rivers that flow northward to the Tanana River. Twenty million years ago the Alaska Range had not yet begun to rise, and the topography was one of low relief, rapidly shifting stream channels carrying sand and gravel, and forested swamps that accumulated thick deposits of plant remains. At that time, the ancestral rivers of interior Alaska flowed south to the area of today's Cook Inlet instead of west to the Bering Sea as major interior drainages do today.

The view from the porthole of our time machine shows us that the vegetation was a mixture of conifers and broadleaf deciduous trees. This forest may have resembled mixed northern hardwood forests that grow today in the northern Great Lakes region and parts of northeastern Asia. Tree types included spruce, pine, hemlock, fir, water pine, redwood, dawn redwood, and broadleaf types such as elm, oak, birch, poplar, maple, willow and hornbeam. For such diverse forests to grow at high latitudes, global climates must have been significantly warmer than today. Mean annual temperatures in interior Alaska were probably 12 degrees to 18 degrees warmer than they are today.

■ *Late Early Miocene to Early Middle Miocene*

Between 18 milion to 15 million years ago a major global warming event occurred. Evidence for this warming has been found in many parts of the world, both on land and in

Natural badland exposures and coal mining excavations at the Suntrana coal mine near Healy provide a rare opportunity to study great thicknesses of rocks of mostly Neogene age (Miocene and Pliocene). Fossil pollen and spores and leaf fossils from these deposits provide important clues to the history of vegetation changes in interior Alaska. (Thomas Ager)

Winter sediment-coring operations at Zagoskin Lake on St. Michael Island in Norton Sound resulted in recovery of a 49-foot-long core spanning more than 40,000 years of history. Lakes such as this occupy several volcanic craters on the island. Crater lakes provide perfect sediment traps that preserve long histories of environmental changes. (Jerry Austin, courtesy of Thomas Ager)

the ocean basins. The reasons for this major warming are uncertain, but they may involve changes in ocean circulation patterns, and a possible increase in atmospheric carbon dioxide. What is clear from the fossil evidence is that ecosystems on land and in the oceans were dramatically influenced by this largest-magnitude warming event of the past 24 million years or more.

In Alaska there are several known sites where fossils of land plants dating to that major warming can be found. We will briefly consider three of these early Middle Miocene sites, which represent a broad latitudinal range from north of the Arctic Circle to the southern coast. One of the sites is in the upper Ramparts area of the Porcupine River. At that locality, ancient lava flows of early Middle Miocene age buried plant remains of forests that are now exposed in the canyon's upper walls. Logs, stumps and thick peat deposits are exposed under and between individual lava flows.

Well-preserved plant fossils (wood, cones, leaves, pollen and spores) demonstrate that forests containing maple, walnut, chestnut, holly, oak, basswood, redwood, dawn redwood, pine, spruce, birch, alder and many other tree types grew in northern interior Alaska during the early Middle Miocene. Most of the tree types found as fossils in rocks of early Middle Miocene age do not grow in or near Alaska today. In sharp contrast to the early Middle Miocene forests, the modern vegetation of the Porcupine River valley consists of only a few types of trees and large shrubs such as spruce, paper birch, aspen, balsam poplar, willows and alders.

Many of the tree types represented in the early Middle Miocene deposits at the upper Ramparts require climatic conditions dramatically different from what exists today. Summers were warmer and winters much milder, and growing season precipitation had to have been much heavier.

Another site of about the same age is at Suntrana, an abandoned coal mining site near Healy, where we previously encountered fossil evidence from somewhat older, early Miocene deposits of the Healy Creek Formation. Rock layers representing ancient lake and river sediments and coal swamp deposits of middle to late Miocene age are exposed in natural badlands and coal mining excavations.

Fossil pollen and fossil leaves from these deposits show that temperate forests with many types of broadleaf deciduous trees grew in the area where the Alaska Range is now. Fossil leaves from the lower part of the Suntrana Formation (Middle Miocene age) represent many of the same tree types that were found farther north in the Porcupine River canyon, such as maple, sweetgum, elm, dawn redwood, basswood, chinese walnut and oak, as well as pine, spruce, hemlock and larch.

A third site with a similar assemblage of leaf fossils and pollen of about the same age is near Seldovia Point, along the southern coast of Kachemak Bay in southern Cook Inlet. The close similarities of fossil assemblages from Seldovia Point, Suntrana and the upper Ramparts of the Porcupine River indicate that temperate broadleaf forests covered at least 9 degrees of latitude in Alaska during the early Middle Miocene. These forests were most similar to species-rich types of northern mixed hardwood forests that now grow in the southern Great Lakes region and parts of northeastern Asia, such as northern Japan, Korea and northeastern China.

Similar assemblages of plant fossils of early Middle Miocene age have now been found at localities extending from northeast Asia as far south as Japan, in the Russian Far East, Alaska and the Pacific Northwest. This suggests that temperate forests of surprisingly similar composition, even to the species level in many cases, occupied much of that large region during the early Middle Miocene warming event. For this to happen, continuous land or closely spaced islands must have existed in the region now under the Bering Sea. Limited fossil evidence from the Aleutian Islands suggests that forests of conifers and broadleaf trees existed in those islands during the Miocene, whereas today they are devoid of forests.

■ *Middle to Late Miocene*

The next stop on our journey through time takes us into a long period of global cooling that began about 14.5 million years ago. By about 12.5 million years ago, many of the temperate trees widespread during the early Middle Miocene were eliminated from the landscape, probably because of cooler summers. Most deciduous trees with broad leaves were replaced by evergreens and deciduous conifers that were better adapted to cooler climates.

The cooling trend continued through the late Miocene, but was interrupted several times by shorter-duration reversals to warmer climates. These oscillations occurred about 11 million to 10.5 million, and 6.5 million years ago, according to evidence from marine sediments in the North Pacific, and never approached the magnitude of the early Middle Miocene warming event.

During the late Middle Miocene and late Miocene, only broadleaf trees and shrubs such as willow, alder, poplar, aspen and hazelnut, along with conifers, were able to tolerate the cooler climates. Willow, poplar and alder diversified into a number of new species during the late Miocene, perhaps in response to many ecological niches that opened up as a result of the loss of broadleaf temperate tree and shrub species.

One noteworthy locality where late Miocene plant fossils are preserved is in bluffs along the Tatlanika River in the Alaska Range. A 25-foot-thick accumulation of volcanic ash buried a conifer-dominated forest more than 8.3 million years ago, and carbonized tree trunks are preserved upright within the ash deposit. At the base of the ash deposit is an ancient forest floor, preserved by rapid burial in dense volcanic ash. The plant fossils that have been identified from this layer are twigs, needles, cones and leaves that came from trees such as fir, pine, larch, hemlock and poplar, as well as shrubs such as willow, alder and spiraea. Local forest vegetation at the site today includes far fewer tree and shrub types: spruce, paper birch, aspen, balsam poplar, alder and willow.

Fossil pollen assemblages from this site and others in the Alaska Range, such as the Lignite Creek Formation and the Grubstake Formation, show additional evidence of long-term global cooling and a gradual elimination of tree types during the late Middle Miocene to late Miocene. By the end of the Miocene, even some of the conifers, such as redwoods, dawn redwoods and water pine, had disappeared from Alaska. Their disappearance may have resulted from cooler summers.

A well-preserved tree stump of an ancient redwood, still rooted in peat from what was once the forest floor, was partly charred at the time of burial by the overlying lava flow, now seen as the blocky gray rock layers above the stump in the Porcupine River canyon. This 15-million-year-old stump has not been fossilized, but is still burnable wood. Dr. James White of the Geological Survey of Canada stands to the right of the stump. (Thomas Ager)

Other late Miocene fossil sites occur along the coastal bluffs of the southern Kenai Peninsula. Well-preserved leaf fossils of willow and alder, and conifer needles can be found in the beach-front bluffs and in rock slabs that have fallen to the beach.

Leaf fossils from early Middle Miocene age occur in a few zones within rock layers of the Suntrana Formation near Healy. These fossils indicate that a forest rich in temperate-climate tree species grew in the Interior during a time of major global warming about 16 million to 15 million years ago. (Thomas Ager)

A time-machine flight about 8 million years ago would reveal dark conifer forests rich in species, interspersed with stands of poplar, willow, birch and alder covering hundreds of thousands of square miles of interior Alaska. In what is now the Cook Inlet area, however, the conifers were far less abundant than stands of cottonwood, willow and alder. At that time no marine embayment had yet developed at the site of Cook Inlet.

By 6.6 million to 6.4 million years ago in the late Miocene, a minor warming event occurred that permitted development of pine- and birch-dominated forests in interior Alaska. Some of our fossil evidence comes from ancient lake deposits now exposed in riverside bluffs of the Porcupine River. At this locality, between Fishhook Bend and Canyon Village, many samples of lake muds were collected to extract pollen and other plant fossils for study. The pollen evidence shows that along with pine and birch, spruce, larch, alder and willow were significant elements of the regional vegetation. Also present were at least small amounts of hemlock, fir, Douglas fir and hazelnut. Hemlock and fir now grow in southern coastal Alaska, but Douglas fir and hazelnut no longer grow in the state.

Aquatic plants represented by fossils at the Canyon Village locality include many types that grow in modern lakes and ponds, such as water lily (Nuphar, Nymphaea) and pondweed (Potamogeton), but which also included water chestnut (Trapa), which no longer is native to North America but grows today in Asia.

At the end of the Miocene another significant climatic oscillation occurred, another cooling event. This global cooling resulted in the expansion of glacial ice in Antarctica. When continental-scale glaciers grow, sea level drops. Some of the water once held in the oceans falls as snow on glaciers and remains stored there for thousands of years. As a result, global sea levels can be lowered. In Alaska, a combination of global cooling and the rise of coastal mountains in southern Alaska resulted in the development of glaciers at least locally, and possibly in higher mountain ranges such as the Wrangell Mountains and the Brooks Range.

We do not yet fully understand what influence that latest Miocene cooling event had on Alaska vegetation, but the information we do have suggests that spruce trees began to increase in abundance, while other conifers such as pine became less abundant.

■ *Pliocene*

The Pliocene began 5.2 million years ago. Few Piocene-age sites found in Alaska include fossil floras that have been adequately studied and dated. One site where we have preliminary studies of fossil pollen assemblages is at McCallum Creek on the southern flank of the Alaska Range near Gulkana Glacier. A dated volcanic ash found there indicates that at least part of the deposit is of earliest Pliocene age, about 5 million years old. Those deposits contain pollen of spruce, pine, birch, larch, alder and a few herbaceous plants. Spruce pollen is more abundant than pine at McCallum Creek, which suggests that the cooling event of the latest Miocene may have continued into the earliest Pliocene.

Marine sediment and fossil records show that global warming events occurred during the middle part of the Pliocene, between about 4.3 million to 3 million years ago. These warm intervals were minor events compared with the major warming of the early Middle Miocene, but they may represent warmer global climates than have existed at any time since.

One of the most interesting Pliocene-age sites being investigated is called Lost Chicken, because the deposits are exposed in placer gold mining excavations at the Lost Chicken Hill Mine in eastern interior Alaska.

The Lost Chicken site contains well-preserved tree stumps, cones and needles of an extinct species of larch, and the plant fossils from the deposits represent spruce, birch, larch, pine, fir and alder. Small amounts of pollen of herbaceous plants such as grass and sedge are present. Pine fossils include a five-needle pine that appears to be related to the modern Korean white pine.

The mid-Pliocene vegetation at Lost Chicken was not drastically different from what grows in interior Alaska today. But even in these relatively young deposits, tree, shrub and herb types represent some extinct plant species or plants that no longer grow as far north as Alaska. The climate was somewhat

ABOVE: *Coal-bearing rocks of late Miocene to Pliocene age are exposed in coastal bluffs of southern Kenai Peninsula from Kachemak Bay north to the Clam Gulch area. Fossil leaves and pollen and spores from these deposits provide evidence of global cooling during the late Miocene. Volcanic ash layers preserved within coal layers can be dated to provide more precise age control for interpreting fossil assemblages. (Thomas Ager)*

RIGHT: *Fossil conifer needles and willow leaves of late Miocene age were found in coal-bearing rocks from coastal bluffs near Homer. (Thomas Ager)*

Pliocene-age river sediments exposed in Lost Chicken Hill placer gold mine near Chicken in the eastern Interior contain ancient layers of forest litter and stumps, twigs, cones and needles of an extinct species of larch. Dr. John Matthews, Jr. from the Geological Survey of Canada stands below a layer of preserved tree stumps of larch. (Thomas Ager)

warmer, with a mean annual temperature slightly above freezing. Permafrost had not yet developed in interior Alaska.

■ Late Pliocene and Pleistocene

The next stop with our time machine takes us to the late Pliocene and early Pleistocene. During this time, the number of tree and shrub types that could survive the increasingly cold climate declined. Trees such as pine, hemlock and fir disappeared from most of Alaska; some species of those types may have survived in coastal regions of southern Alaska, at least during times when glaciers did not cover the region.

Between 3 million to perhaps 1 million years ago, spruce, larch, birch, alder, willow, poplar and herbaceous plants became dominant. Although glaciers developed in some mountain ranges during the latest Miocene, the first large-scale glaciation occurred about 2.5 million years ago, during the late Pliocene. Our current understanding of what happened during that early glacial episode and during most of the Pleistocene ice ages and the intervening warm periods is limited. What we can say is that there were many intervals of glacial climate during the past 2.5 million years, and those intervals lasted about 100,000 years.

Warmer periods (interglacials) usually lasted about 10,000 to 20,000 years. This long period of oscillating climates dramatically influenced Alaska vegetation. Some fossil plant evidence suggests that tundra or tundralike plant communities began to develop in northern Alaska during the latter part of the late Pliocene. Such communities may have developed earlier and in other areas of Alaska, such as in the high mountain ranges, but we do not have fossil evidence to prove it. Alpine floras are rarely preserved as fossils.

During the late Pliocene and early Pleistocene glacial intervals, perennial snow and ice covered up to 50 percent of Alaska; in the late Pleistocene, they covered about 30 percent. Today glaciers cover only about 3 percent of the land surface. During glaciations, large areas of continental shelf in the Bering and Chukchi seas and along the northern coast were exposed.

Surprisingly, large areas of interior and northern Alaska escaped being buried by glacial ice. The development of high mountain ranges across southern Alaska during the Pliocene and Pleistocene prevented large amounts of moisture from penetrating into the interior. The dry continental climate of the interior had too little snowfall to produce large glaciers.

These unglaciated lands were part of a continuous unglaciated land mass called Beringia that extended from the Yukon

Territory in Canada to the east, and west into eastern Russia across the Bering Land Bridge. This land connection permitted exchange of plants and animals between northeastern Asia and northwestern North America during long intervals of the past 2.5 million years. But what plant life was there to exchange? During glacial times, the areas of Alaska beyond the limits of glacial ice appear to have had a different climate than exists today. There was less precipitation in interior regions, and winters were probably colder.

Much of what we now know about Alaska's vegetation during the last glacial interval and during the Holocene interglacial that followed comes from pollen, spores, seeds and other plant macrofossils preserved in lake sediments and peat bogs. Much of this evidence has been obtained by collecting sediment cores from lake bottoms.

Fossil evidence suggests that during the most severe intervals, such as during the most recent glacial advances that began about 26,000 years ago and ended between 14,000 and 10,000 years ago, forests vanished from Alaska. A few species of trees may have survived in Beringia, but if they did, they were probably restricted to marginal habitats, possibly near flowing springs or hot springs. In such habitats only the hardiest tree types such as balsam poplar and aspen, and species of larger shrub willow are likely to have survived. There is not yet compelling evidence that spruce trees survived in Alaska during the last glaciation.

Pollen assemblages that accumulated in lake

Samples of sediments were collected from deposits of ancient lake beds of late Miocene age exposed in bluffs near Fishhook Bend along the Porcupine River. These deposits preserve pollen and spores, diatoms, plant macrofossils and volcanic ash. A thin layer of volcanic ash in the lake beds was isotopically dated and found to be about 6.5 million years old. (Thomas Ager)

sediments during the last glaciation contain essentially no pollen of trees, and even pollen of woody shrubs is rare. Pollen of grass, wormwood (sage), aster, sedge and a variety of herbaceous flowering plants (pink family, crowfoot family, Jacob's ladder) suggest tundralike vegetation extended across Beringia. Shrubs were mostly willow, while blueberry, crowberry and cranberry were present but not widespread. At least one species of alder appears to have endured the last ice age in lowland areas of northwestern Alaska.

Was the tundralike vegetation similar to modern tundra communities of Alaska? It is highly likely that some of it was similar to tundra vegetation that can be found today in alpine zones, interspersed with open patches of barren soils or shattered rock, with relatively few mossy habitats where a thick organic mat could cover the soil.

However, persuasive arguments have been made by vertebrate paleontologists who think that the full-glacial vegetation of Beringia had to be different from modern tundra communities to support the diverse assemblage of large mammals that inhabited the region. Most large grazing mammals require high-quality forage in abundance for many months of the year. Modern tundra communities produce forage that deteriorates in nutritional quality by late summer.

The ice-age vegetation of Beringia may have included some plants that now grow in the grasslands of the northern Great Plains or in eastern Russia steppe-tundra; it may also have included tundra plants now found in Alaska. A flight over Alaska 18,000 years ago during midsummer would have shown large glaciers in the mountains, and a treeless, or nearly treeless, interior covered by patches of tundra, grasses and bare silty soils.

By about 13,000 years ago, a shrubbier vegetation that included dwarf birch, willow and blueberry began to supplant the herbaceous tundra or steppe-tundra of full-glacial times. The spread of shrub tundra coincides with the time of retreating glaciers, and was probably a response to warmer climates and deeper snowfalls.

Balsam poplar, perhaps aspen, and larger willow shrubs began to spread across Beringia about 11,000 years ago, probably in response to warmer summers. The most convincingly dated early man sites, ca. 12,000 to 10,000 years old, indicate that human populations were spreading throughout the region during this time when shrub birch tundra, willow thickets and then poplar and aspen groves were rapidly replacing herbaceous tundra or steppe-tundra, and when the huge glacier systems were rapidly waning, discharging great volumes of meltwater into the river systems. With dramatic environmental changes, the remnants

Basaltic lava flows 14 million to 17 million years old are exposed in steep cliffs along the upper Ramparts of the Porcupine River canyon. These lava flows covered ancient forests, preserving logs and stumps, thick peat layers, pond deposits, seeds and occasional leaf fossils. Fossils from these beds provide new evidence that temperate forests grew north of the Arctic Circle about 16 million to 15 million years ago. (Thomas Ager)

of large-mammal populations may have found it increasingly difficult to find suitable forage.

■ *Holocene*

The Holocene, which began 10,000 years ago, is an interglacial that will probably be followed by a long period of glacial climates thousands of years from now. The history of vegetation change during the Holocene is of particular interest, because it allows us to see how and when modern vegetation patterns came into being. Human populations that entered Alaska shortly before the beginning of the Holocene had to adapt to significant changes in the landscape, climate, vegetation and mammal populations that followed deglaciation.

Our time-travels show that during the early Holocene, spruce trees began to spread across lowlands where only tundra, shrublands and scattered poplar and aspen stands had existed previously. Spruce pollen appears in many interior Alaska sites about 9,400 to 9,000 years ago. Spruce appears to have spread into interior Alaska from northwestern Canada.

From interior Alaska, spruce advanced southward, westward and northwestward, reaching some areas where they now grow within the past few thousand years. The first type of spruce to colonize interior Alaska was white spruce, followed by black spruce about 1,000 years or more later. Alder shrubs spread across large areas of Alaska beginning about 8,000 years ago.

The extensive areas of southern Alaska that were covered by snow and ice were gradually forested by trees and shrubs derived from the coastal forests of the Pacific Northwest, portions of which had survived in parts of Washington, Oregon and northern California during the last ice age. That gradual westward advancement of coastal forest vegetation is underway in southcentral and southwestern Alaska.

Modern tundra vegetation communities of Alaska include many species of herbaceous plants and low woody plants such as willows that existed in Beringia during past ice ages. However, some plants that were widespread during the last glaciation may now be restricted in distribution, and some may now be extinct or no longer growing in Alaska. Some plants, especially mosses, that are common in Alaska tundra vegetation today are clearly far more abundant now than they were during the ice ages. Tundra vegetation is highly varied in species composition from place to place, and we are discovering that its history is surprisingly complex. The plants comprising tundra vegetation are a combination of recently evolved herbaceous species as well as relict species inherited from landscapes that preceded the ice ages. It is an important part of the Alaska ecological history that we have only begun to understand. ■

BELOW: *Less severe climate in western coastal Alaska permits development of a richer assemblage of shrubs and flowering tundra plants than is found on the North Slope. These flowers brighten a hillside near Tununak. (Thomas Ager)*

RIGHT: *Two thick volcanic ash layers of late Miocene age are exposed in banks of the Tatlanika River. The lower ash, shown here, is about 25 feet thick, and was deposited by a volcanic eruption more than 8.3 million years ago. The ash preserved standing tree trunks, shown here as black vertical structures at upper center, and a layer of forest litter at the base of the ash. (Thomas Ager)*

Pleistocene Mammals

By Paul E. Matheus

Editor's note: *Paul Matheus is completing a doctorate in paleobiology at the University of Alaska's Institute of Arctic Biology. His research deals with the evolutionary ecology of Pleistocene mammals, particularly carnivores.*

Matheus would like to thank his mentor and graduate advisor, R. Dale Guthrie, who inspired many of the images described in this chapter.

For anyone who has ever wandered through an Alaska black spruce forest or trekked across boggy tussocks, the taiga and tundra appear to be timeless and resilient. It is also easy to imagine that the familiar mammals inhabiting these landscapes – caribou, moose, bears, hares, fox, wolves, squirrels, voles – must also be ageless components.

But as you have been reading, Alaska's modern taiga and tundra ecosystems and many of the animals which inhabit them are young, only about 10,000 years old. Although many of Alaska's mammals are relative newcomers to Alaska, a few are remnants from what used to be a spectacular and more diverse mammal fauna of ice-age Alaska. Only recently lost to extinction are great Alaska beasts such as woolly mammoths, wild ponies, steppe bison, saber-toothed cats, steppe lions and giant short-faced bears. Unlike the arcane, distant dinosaurs that were long gone before the first primate ever swung through a tree, some of these ice-age mammals may well have slept under the midnight sun among people indistinguishable from you and me.

In this chapter, we will explore how physical and biological processes shaped the ecology of Alaska's mammal community through the late Pleistocene and early Holocene, setting the stage for our modern ecosystems. The story is one of finicky climates and luck. Some mammals gambled on biological specializations and lost, while others survived by simply being in the right place at the right time.

■ *The Pleistocene Climate*

You have read how the Pleistocene represents a time of worldwide climatic cooling and onset of glacial ice ages. The Pleistocene was not one continuous cold period. Instead, glacial periods came and went throughout the Pleistocene, interposed by warmer interglacial periods that in some ways resembled today's climate.

The Pleistocene's last great ice age, the Wisconsinan Glaciation, is the setting for most of this chapter's discussion because more than 90 percent of Alaska's prehistoric mammal fossils date to that period. The Wisconsinan ice age began around 100,000 years ago and lasted almost to the end of the Pleistocene, 10,000 years ago. We know little about Alaska mammals from earlier Pleistocene times because their fossils are rare. Even so, for most of the Pleistocene, Alaska was an important bio-geographic crossroads in the world. Most animals that traveled between the Old World of Europe and Asia to the New World of North America had to come through Alaska, yet they left almost no trace of their presence here prior to the Wisconsinan.

Even the Wisconsinan was not one long cold spell. Its glacial climate was interrupted between 30,000 and 60,000 years ago, when warmer, wetter conditions and forests temporarily returned to Alaska. The last glacial surge of the Wisconsinan began about 24,000 years ago and peaked around 18,000 years ago. By 14,000 years ago the glaciers were rapidly receding, and by 10,000 years ago the great continental ice masses were all but gone.

The previous chapter described how throughout the ice ages vast stretches of interior, western and northern Alaska remained free of glaciers. Alaska's glacial-free areas were part of a larger biological refugium called Beringia. With the Bering Land Bridge high and dry, Beringia extended from near the MacKenzie River in Canada's Yukon Territory all the way to the Lena River in eastern Siberia. Glaciers grew only locally in Beringia's higher peaks like the Brooks Range and Kuskokwim Mountains. However, glaciers in the southern coastal mountains and Alaska Range engulfed most of southern Alaska.

Precipitation, as much as cold temperatures, influences glacier growth. Glaciers never proliferated in the Interior because the Alaska Range and coastal mountains blocked oceanic moisture from reaching this far north. Even in today's comparatively wetter climate, Fairbanks still only receives around 11 inches of total annual precipitation.

■ *A Productive Grassland in the Far North*

The dryness that prevented glaciers forming in the Interior also led to development of a cold, arid and treeless grassland called a steppe. The steppe extended throughout Beringia during peak glacial times. During dry cool phases of the Wisconsinan, the steppe stretched from the Yukon Territory through Alaska, northern Asia and Europe all the way to England, which was attached to Europe during low sea levels. While this northern landscape may sound bleak, it was actually a uniquely structured, immense ecosystem that supported a diversity of grazing and carnivorous creatures, the famous ice-age mammals. Borrowing from Beringia's most famous inhabitant, the elephantlike mammoth, paleontologists refer to this menagerie as the Mammoth Fauna. The ecosystem as a whole has been dubbed the Mammoth Steppe.

To visualize animals on the Mammoth Steppe, we first need to look closer at its environment and plants. In the past it has been described as a steppe-tundra or tundra grassland, but the word tundra can be misleading. True tundras are more than just cold, treeless ecosystems. Tundras are usually poorly drained with slow rates of organic decomposition and recycling. Tundra plants are slow-growing species with special adaptations and conservative growth strategies for cold climates. Many bryophytes – mosses,

With tusks up to 12 feet long and weighing more than 200 pounds, the woolly mammoth was the granddaddy and namesake of the Mammoth Steppe in Alaska. Unlike modern elephants, the woolly mammoth was almost exclusively a grazer, grabbing clumps of grass with the split, handlike tip of its trunk. Mammoth mummies found in Alaska and Siberia have taught scientists about the life and ecology of these beasts. While accounting for only 5 percent of Pleistocene fossils found in Alaska, mammoths constituted almost a third of the ecosystem's mammalian biomass, because of their size. (Natural History Museum, London)

Alaska's Prehistoric Mammals

EARLY AND MIDDLE PLEISTOCENE MAMMALS (Pre-Wisconsinan Age)

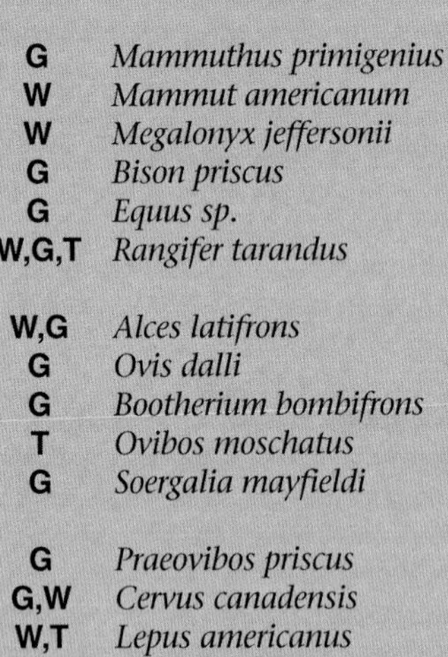

Herbivores

Common Name	Habitat	Scientific Name
Woolly Mammoth	G	*Mammuthus primigenius*
American Mastodon	W	*Mammut americanum*
Giant Ground Sloth	W	*Megalonyx jeffersonii*
Steppe Bison	G	*Bison priscus*
Large Horse	G	*Equus sp.*
Caribou	W,G,T	*Rangifer tarandus*
Broad-Fronted Moose	W,G	*Alces latifrons*
Dall Sheep	G	*Ovis dalli*
Helmeted Musk Ox	G	*Bootherium bombifrons*
Tundra Musk Ox	T	*Ovibos moschatus*
Soergel's Musk Ox	G	*Soergalia mayfieldi*
Staudinger's Musk Ox	G	*Praeovibos priscus*
Elk	G,W	*Cervus canadensis*
Snowshoe Hare	W,T	*Lepus americanus*
?Collared Pika	G,T	*Ochotona collaris*
Giant Pika	G,T	*Ochotona whartoni*
Marmot	G,T	*Marmota sp.*
Arctic Ground Squirrel	G,T	*Spermophilus perryii*
Beaver	A	*Castor canadensis*
?Giant Beaver	A	*Castoroides ohioensis*
Muskrat	A	*Ondatra zibethicus*
?Porcupine	W	*Erithizon dorsatum*
Brown Lemming	T,G	*Lemmus sibiricus*
Collared Lemming	T,G	*Dicrostonyx sp.*
Bog Lemming	T,A,G	*Synaptomys sp.*
Meadow Mouse	G,W	*Pliomys deeringensis*
?Red-Backed Vole	T,W,G	*Clethrionomys gapperi*
Other Voles	W,G,T,A	*Microtus spp.*
?Shrews	W,G,T,A	*Sorex spp.*

Carnivores

Common Name	Habitat	Scientific Name
Steppe Lion	G	*Panthera leo*
Sabertooth	W,G	*Homotherium sp.*
?Cheetah	G	*Acinonyx trumani*
?Short-Faced Bear	G	*Arctodus simus*
Bear (Black?)	W	*Ursus sp.*
Wolf	W,G,T	*Canis lupus*
Dhole (Wild Canid)	W	*Cuon sp.*
?Coyote	W,G	*Canis latrans*
?Wolverine	W,G,T	*Gulo gulo*
?Lynx	W	*Lynx canadensis*
Fox	W,G,T	*Vulpes* or *Alopex*
Spotted Skunk	W	*Spilogale sp.*
Mink	A	*Mustela vison*
Pine Marten	W	*Martes americana*
Ermine	T,W	*Mustela erminea*
?Least Weasel	W,G,T	*Mustela rixosa*

LATE PLEISTOCENE MAMMALS (Wisconsinan Age)

Herbivores

Common Name	Habitat	Scientific Name
Woolly Mammoth	M	*Mammuthus primigenius*
American Mastodon	W	*Mammut americanum*
Giant Ground Sloth	W	*Megalonyx jeffersonii*
Steppe Bison	M	*Bison priscus*
Alaska Wild Ass	M	*Equus hemionus*
Alaska Pony	M	*Equus caballus*
Saiga Antelope	M	*Saiga tatarica*
Caribou	M	*Rangifer tarandus*
Helmeted Musk Ox	M	*Bootherium bombifrons*
Tundra Musk Ox	M	*Ovibos moschatus*
Camel	M	*Camelops sp.*
Broad-Fronted Moose	M	*Alces latifrons*
Dall Sheep	M	*Ovis dalli*
Elk	M	*Cervus canadensis*
Snowshoe Hare	M	*Lepus americanus*
Collared Pika	M	*Ochotona collaris*
Marmot	M	*Marmota sp.*
Arctic Ground Squirrel	M	*Spermophilus perryii*
Beaver	M	*Castor canadensis*
Giant Beaver	W	*Castoroides ohioensis*
Muskrat	W	*Ondatra zibethicus*
Porcupine	W	*Erithizon dorsatum*
?Red Squirrel	W	*Tamiasciurus hudsonicus*
Brown Lemming	M	*Lemmus sibiricus*
Collared Lemming	M	*Dicrostonyx torquatus*
Bog Lemming	M	*Synaptomys borealis*
Red-Backed Vole	M	*Clethrionomys gapperi*
Meadow Vole	M	*Microtus pennsylvanicus*
Singing Vole	M	*Microtus miurus*
Tundra Vole	M	*Microtus oeconomus*
Arctic Shrew	M	*Sorex arcticus*

Carnivores

Common Name	Habitat	Scientific Name
Steppe Lion	M	*Panthera leo*
Sabertooth (Scimitar)	M	*Homotherium serum*
?Cheetah	M	*Acinonyx trumani*
Short-Faced Bear	M	*Arctodus simus*
Grizzly (Brown) Bear	M	*Ursus arctos*
Wolf	M	*Canis lupus*
?Coyote	M	*Canis latrans*
Wolverine	M	*Gulo gulo*
Lynx	M	*Lynx canadensis*
Arctic Fox	M	*Alopex lagopus*
Red Fox	M	*Vulpes vulpes*
Badger	M	*Taxidea taxus*
Black-footed Ferret	M	*Mustela nigripes*
Spotted Skunk	W	*Spilogale sp.*
Mink	M	*Mustela vison*
Pine Marten	W	*Martes americana*
Ermine	M	*Mustela erminea*
?Least Weasel	M	*Mustela rixosa*

—And undoubtedly many more not yet discovered.

W = woodland species
T = tundra species
G = grassland/steppe species
A = aquatic (wetland) species
M = mammoth steppe species
? = questionable, but likely

sphagnum and lichens – grow on the tundra. Typical examples of more advanced tundra plants are evergreen forbes, sedges, grasses and birch and willow shrubs.

Steppe environments, in contrast, are driven by different processes. Aridity, more than cold, characterizes steppe lands. Their better drained, exposed and thus warmer soils allow plants to grow deeper roots. Organic material also decomposes at higher rates on the steppe, bringing on more grasses and sedges. Because of their extensive root systems, grassy species can afford to loose their above-ground growth. That spells grazing. Steppe lands are much more productive for grazing than tundra, because almost all green growth of the steppe is edible. Most tundra foliage contains toxic chemicals, rendering the plants inedible; the little remaining non-toxic, edible growth supports fewer animals.

Yet the Mammoth Steppe was more than an endless grassland. Forbes, willows and other shrubs, even sagebrush, grew among the grasses and sedges, depending on the land's elevation, slope, drainage and soil type. Even true tundra was undoubtedly found on some parts of the Mammoth Steppe. As a result, the steppe was a heterogeneous landscape with a variety of habitats.

The mammals of Alaska's Pleistocene steppe exhibited special adaptations for life in this cold, unique landscape. Based on their abundance as fossils, three grazing mammals – mammoth, bison and horse – dominated the Mammoth Steppe. However, many other beasts played important roles to balance this ecosystem.

■ *Ice-Age Beasts: Grazing Machines*

The familiar woolly mammoth, Alaska's most colossal ice-age mammal, looked like a short, shaggy elephant with droopy hind quarters and a domed head topped by a comical toupeelike tuft of hair. Its back end held a tiny, hairy tail, while its front waved a unique appendage, a trunk split at the tip, like the thumb and fingers of a hand. As a grazer, the mammoth used its trunk "hand" to grab bunches of grass to eat. It may have used the same trick to put snow in its mouth for water. Mammoths sported tusks of ivory, up to 12 feet long, which both sexes used to display social status and fitness. Elephants today use their tusks in much the same way. Like their modern relatives, mammoths backed up displays of their threatening weapons with lethal demonstrations. Many fossilized mammoth tusks are broken or worn, testaments perhaps to ancient feuds.

Mammoths were extremely successful in Alaska, despite their large size and demands for

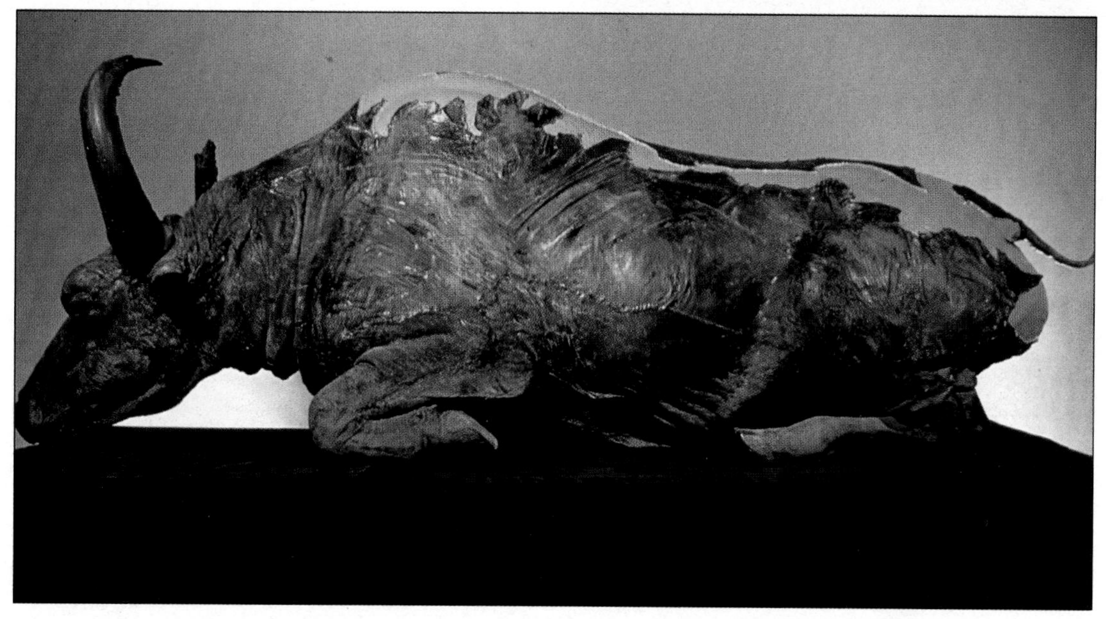

Frozen and dehydrated mummies like these from Alaska's frozen sediments have taught us a great deal about the environment and ecology of Pleistocene Alaska. This ground squirrel (right) never roused from its deep hibernation 33,000 years ago. Plants lining its undisturbed burrow help reconstruct the vegetation on the surface. Blue Babe (below) is a famous steppe bison mummy recovered from a mine in the hills north of Fairbanks. Blue Babe was so well-preserved that its flesh was still red. Its hide was preserved and is now on display at the University of Alaska Museum in Fairbanks. (Both, University of Alaska Museum, courtesy of Paul Matheus)

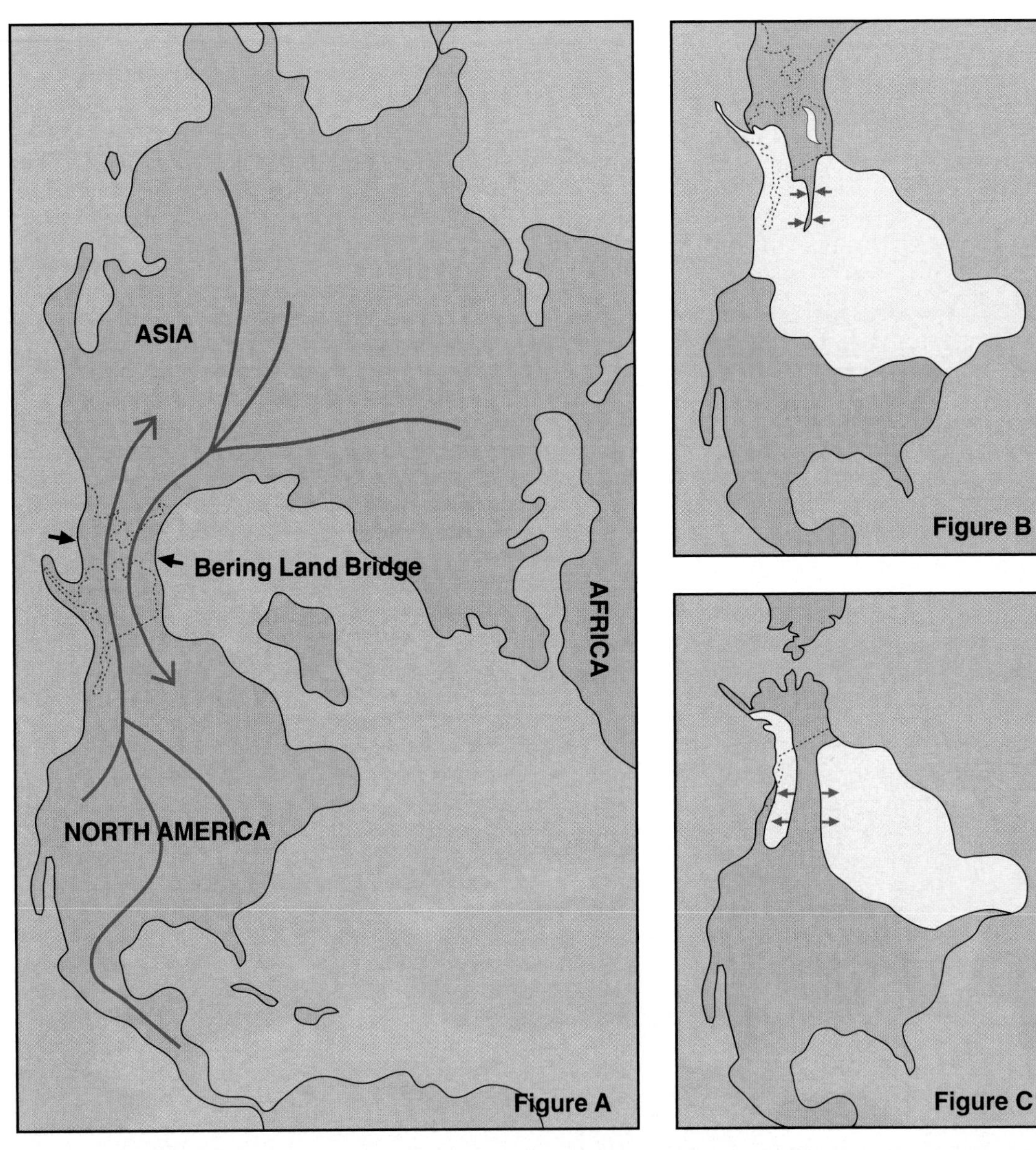

The First Alaska Highway. *During most of the Cenozoic prior to the Pleistocene, Alaska was the only land route for dispersal of plants and animals between the Old World and New World (figure A). About 3 million years ago the Bering Land Bridge became inundated, cutting off Alaska from Asia and breaking this important link. Later, during cold intervals of the Pleistocene, so much of the Earth's water became incorporated into glacial ice that sea levels dropped up to 300 or more feet and temporarily re-exposed the Bering Land Bridge. At the same time, Alaska became isolated from the rest of North America by coalescing ice sheets to the southeast over Canada, and Alaska's mammals were more Asiatic (figure B). When Pleistocene climates swung to the warm side, glaciers melted, sea-level rose, and the Land Bridge was cut — opening up Alaska to the rest of North America and reuniting them biologically (figure C). (Adapted from R. Dale Guthrie; graphic by Kathy Doogan)*

massive amounts of grass. Even though mammoth bones make up only about 5 percent of the total late Pleistocene fossils in Beringia, they constituted around 30 percent of the large mammal biomass.

Mammoths have gained a lot of attention lately as their mummified remains have been found in frozen Pleistocene silts of Alaska and Russia. Unlike the folkloric images, however, these mammoths were not caught frozen in their tracks in blocks of glacial ice. Mammoth mummies formed in rare instances when dead animals were buried soon after death by sediments, while a corresponding rise in the permafrost table froze the sediments and carcasses. The mummification, or drying out, occurred during subsequent millennia as frozen ground slowly sucked moisture out of the carcasses, analogous to what happens with freezer-burned meat. Just imagine what an unwrapped moose steak would look like after being left in a freezer for 30,000 years. For that reason, mummies from Beringia often look like dried and deflated shadows of former creatures.

We have learned a lot about Pleistocene mammals from mummified remains. That is how the mammoth's toupee and handlike trunk were discovered. Mummies from about 10 other species of Pleistocene mammals have been uncovered in Alaska, including steppe bison, musk oxen, horses, moose, caribou, pikas, ground squirrels, hares, voles and lynx.

A famous steppe bison mummy, called Blue Babe because of the blue metallic mineral coating on its hide, was unearthed at a mine just outside of Fairbanks in 1979. Dale Guthrie from the University of Alaska excavated Blue Babe and performed an extraordinary necropsy on the carcass. His careful analysis revealed that Blue Babe was a mature, healthy bull killed in late autumn by two or three steppe lions about 36,000 years ago. The preserved remains of Blue Babe are now on display at the University of Alaska Museum in Fairbanks.

Mummies of steppe bison are rare, but the overwhelming abundance of fossilized bison bones indicates that it was the predominant species on the Mammoth Steppe. Steppe bison are closely related to modern plains bison, but their behavior and ecology were strikingly different.

Steppe bison most likely did not form massive migratory herds like modern plains bison. Today's short-grass prairie of North America is more homogeneous than was the Mammoth Steppe. The prairie grasses grow green and nutritious for a short time each summer, with different large grass pastures ripening to peak nutrition at different times. So to get enough to eat, the plains bison migrate in large herds across the prairie, mowing down each pasture as it turns green.

In contrast, on the Mammoth Steppe, grasses grew in small patches mixed among other plants. The steppe also held more diverse grass and sedge species, which extended the grazing season as each species greened up at different times. The steppe bison tended to gather in small groups that were distributed widely throughout the steppe. These steppe bison moved around in search of the best grass, but the overall makeup of the steppe offered no incentive for them to make long-range seasonal migrations.

The position, size and shape of the neck, hump and legs of steppe bison also indicate that they were not designed to be long-range migrators. Furthermore, their distribution at different fossil localities suggests the bison's primary habitat was on hillsides, rather than

Mammals of Alaska's Mammoth Steppe, Late Pleistocene *(overleaf). Glaciation advanced and retreated many times during the Pleistocene Epoch, 1.8 million years ago to 10,000 years ago. Yet despite the Pleistocene's generally cold climate and repeated ice ages, mammals thrived. During the coldest periods of ice surges, mammals inhabited unglaciated steppe lands that, at various times, stretched through North America, Asia and Europe. These are often collectively called the Mammoth Steppe after the woolly mammoth, one of the Pleistocene's most famous mammals. Intermittently during the Pleistocene, warmer temperatures caused glaciers to retreat and allowed tree growth. The Pleistocene woodlands hosted a different mammal community that included the giant ground sloth and elephantlike mastodon.*

This illustration shows the mammals of Alaska's Mammoth Steppe during one of the last great ice ages, the Wisconsinan Glaciation of the late Pleistocene. Various types of grasses and sedges grew in patches on the Mammoth Steppe, supporting many grazing mammals that, in turn, were prey for large carnivores. Forbes, willow shrubs and sagebrush grew among the grasses, and the mix of plants varied with elevation, moisture and the type of soil. Permafrost in some areas caused the ground to heave and crack in polygons. The moving glaciers crushed rocks and soils into fine particles carried by the wind, causing dust clouds on the horizon. This windblown silt, or loess, accumulated in deep layers and dunes across the land, which helped preserve a fossil record of Alaska's late Pleistocene mammals. (Painting by Tom Stewart)

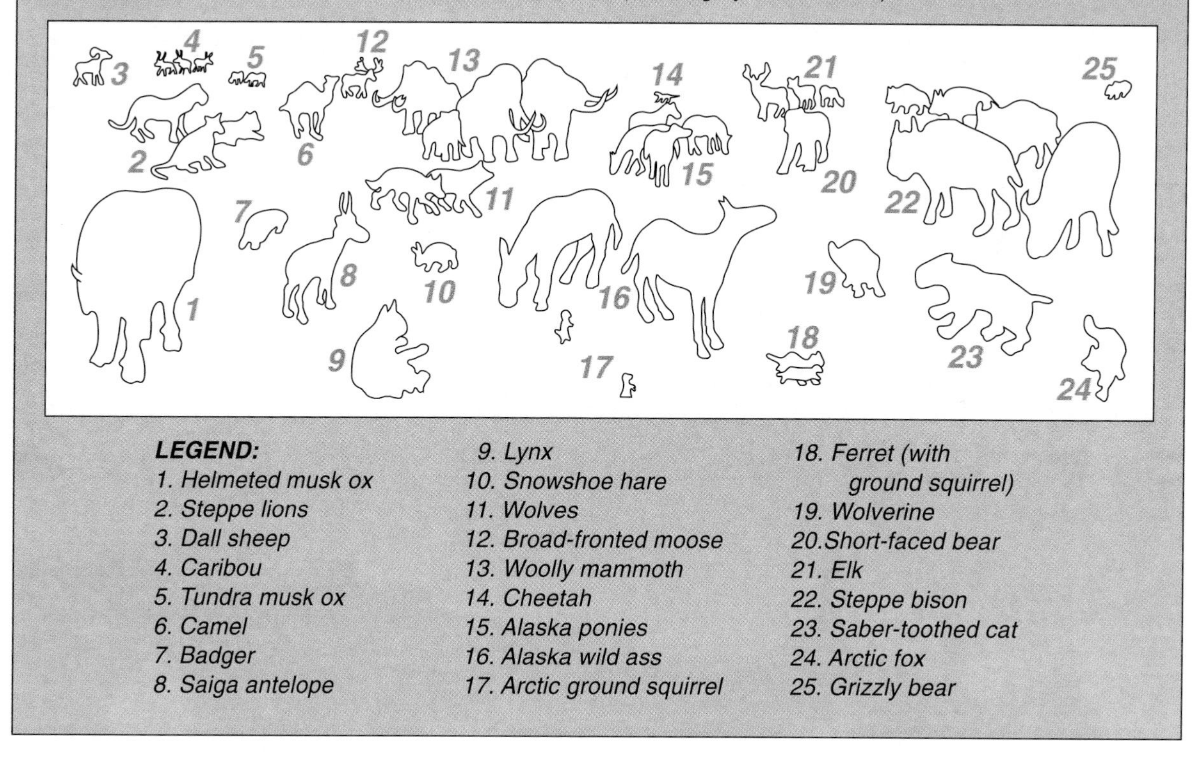

LEGEND:
1. Helmeted musk ox
2. Steppe lions
3. Dall sheep
4. Caribou
5. Tundra musk ox
6. Camel
7. Badger
8. Saiga antelope
9. Lynx
10. Snowshoe hare
11. Wolves
12. Broad-fronted moose
13. Woolly mammoth
14. Cheetah
15. Alaska ponies
16. Alaska wild ass
17. Arctic ground squirrel
18. Ferret (with ground squirrel)
19. Wolverine
20. Short-faced bear
21. Elk
22. Steppe bison
23. Saber-toothed cat
24. Arctic fox
25. Grizzly bear

Not an inhabitant of the Mammoth Steppe, the mastodon was a tree-dependent browser of woody plants and therefore only lived in Alaska during warmer interstades and interglacials when forests and woodlands abounded. Because forest soils do not preserve fossil bones well, mastodon and other interglacial fossils are rare in Alaska. (M. Long, Natural History Museum, London)

the plains or valley bottoms. Even in lower elevation sites, however, bison make up about one third of all Pleistocene mammal fossils.

Bison entered Alaska from Asia about 400,000 years ago and became so successful during the Pleistocene that they competed with and displaced at least two ancient species of musk oxen. However, two other species, the modern musk ox and the more primitive helmeted musk ox, managed to coexist with bison on the Mammoth Steppe. Both were fairly common across the steppes of ice-age Alaska. While musk oxen superficially resemble bison, they are quite different, more closely related to sheep and goats. Various species of musk oxen have lived in Beringia at various times since the late Pliocene. While we do not have a good fossil record of their early history here, much of their evolution may have occurred in Beringia.

The modern musk ox, which is a true Pleistocene relic, withstood the bison onslaught because unlike bison, the musk ox is not a grazing specialist of open grasslands. Musk oxen in Pleistocene Alaska probably frequented the same places as they do today, arctic river valleys where they browse on tender willow shoots and riparian sedges and grasses.

The long-legged helmeted musk ox ate more grasses than its compatriot and may have had more migratory habits. Instead of sticking to riparian and shrubby areas, it also wandered the dry plains where it subsisted on a more grassy diet. Plant fragments scraped from grooves in helmeted musk ox teeth reveal grasses, willow and various shrubs; the plant remains could reflect seasonal changes in the helmeted musk ox's diet. Helmeted musk oxen were widespread in Pleistocene North America, but they became extinct when the dry grassland habitat shrunk below critical levels. They lost out to competitors better suited to using the later Holocene habitats.

After bison, horses were the second most common late Pleistocene mammals. Horses have a long history in Beringia, starting at least with the early Pleistocene. We know little about these early horses, except that they were larger than those of the late Pleistocene.

During the late Pleistocene, two species from the horse family lived on Alaska's Mammoth Steppe. One was a miniature horse. The other was bigger, probably similar in appearance and behavior to today's wild Asian ass. Asses have larger heads, longer ears and smaller hooves than horses. Mummies reveal that the small horse was reddish with black legs, mane and tail. The Alaska wild ass was chestnut-colored with blond tail, mane, belly and flanks. It is unclear how these two similar species coexisted on the Mammoth Steppe, but the fact that they did is testimony to the steppe's productivity and diversity.

Horses and asses have grazing strategies well-adapted to the steppe. They consume large quantities of low-quality grasses. Their teeth can withstand the abrasion inflicted by chewing so much grass, and their guts are similarly adapted. Masses of grass pass quickly through their digestive tracts. However, this high-fiber, high-volume diet lacks certain essential micro-nutrients, which the horses obtain by occasionally browsing on forbes and shrubs. Interestingly, mammoths had a somewhat similar approach to eating.

This foraging strategy reduced the horses' competition with bison on the grasslands. The steppe bison grazed only the greenest patches of grass, and they ate smaller quantities at a time because of their multi-chambered, ruminant stomachs. Grass would enter the first chamber, be regurgitated as moist cud, which the bison would re-chew and swallow for digestion in the other stomach chambers. This allowed the bison to extract the most nutrients from small amounts of high-quality forage. Horses, in contrast, could eat old brown stems as long as they ate a lot and browsed an occasional flower or sagebrush. The broad, grassy flats, valley bottoms and gentle hillsides of the Mammoth Steppe must have been ideal horse habitat.

Using these ecological conclusions, imagine the Mammoth Steppe on a cool, breezy midsummer day. A few bison might be lying on a green hillside, ruminating cud and scanning for predators stalking through sparse short grasses. Below in broad valleys where most of the grasses already were brown, herds of anxious little horses might be grazing and on the move, an occasional head popping up to make sure the two lions in the draw have not crept any closer. If the lions moved in, the fleet ponies would be off in an instant, running to safety before taking up their grazing again without much more than a nervous blink. Across the valley, an old mammoth matriarch would be leading her sisters, aunts and their young on a twice daily trek to the watering hole. Cacophonous trumpeting from their suitors and brothers would rattle the evening's approach in long shadows under a still-bright midnight sun.

While the Alaska steppe would be filled mostly by bison, horses and mammoth, a closer look would reveal other, less abundant animals filling different niches.

Camels might seem an unlikely inhabitant of Alaska, but until about 12,000 years ago, they cruised hillsides around areas like Fairbanks and the North Slope. The habits and ecology of modern camels show that these creatures were perfectly suited to life on the arid Mammoth Steppe – especially when conditions reached the driest extreme during peak glacial episodes. Alaska's camels were probably more related to llamas of South America than to the more familiar Old World camelids. However, the Alaska camel's ecology probably

Meandering rivers throughout interior, western and northern Alaska cut through frozen muck and hills of loess (wind-blown glacial silt from the glaciated Alaska and Brooks ranges deposited on the surrounding land during the Pleistocene). These cutbacks sometimes expose ice wedges, such as shown in the dark area of this bank. Each year during breakup, more bones are eroded from the riverbanks and deposited on shores and bars. It is not unusual while canoeing down these rivers to see the tip of a mammoth tusk protruding from a tall cutbank of loess, or to find a 30,000-year-old bison jaw among the stones on a midriver gravel bar. (Mary E. Edwards)

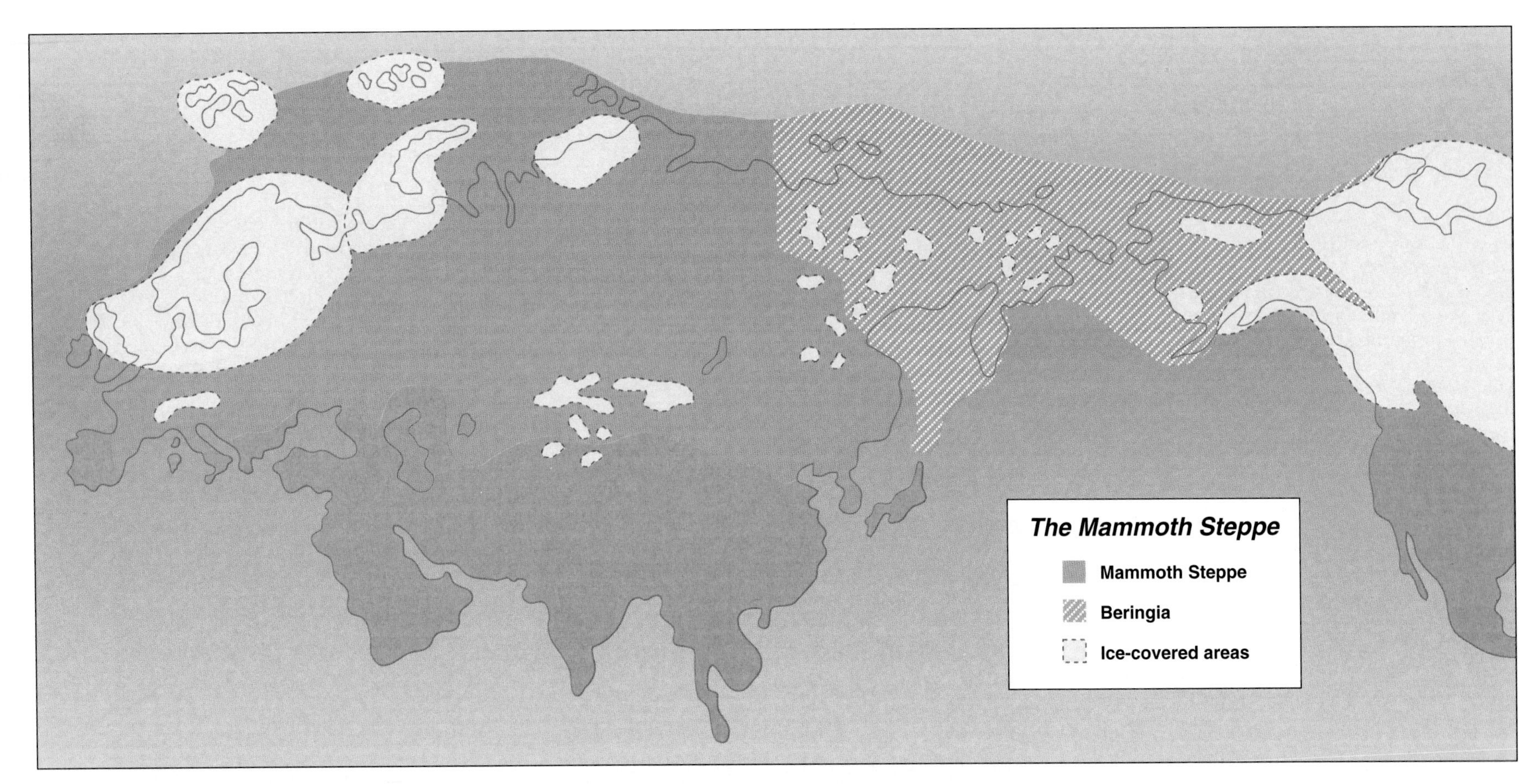

The Mammoth Steppe. *During cold periods of the Pleistocene, Alaska and Beringia were part of an even vaster ecosystem stretching from Canada's Yukon Territory, through Alaska, into Asia and Europe, all the way to England. Named after its most famous inhabitant, this cold and arid grassland supported a greater diversity of mammals than the modern tundra. Dominated by grazing mammals, the three most abundant species were bison, horse and mammoth, but many other fascinating creatures filled niches on this ice-age landscape. (Adapted from R. Dale Guthrie; graphic by Kathy Doogan)*

overlapped that of the modern bactrian (two-humped) camel of the Asian high plateau.

While other herbivores succeeded by occupying a narrow niche, specializing in a foraging style or food type, camels survived on a varied diet. Camels can eat practically any vegetation, often eating what other herbivores will not touch. However, their splayed, soft hooves limit them to smooth, fairly flat ground. They were probably driven out of Alaska by stumps, root wads, bogs and tussocks inherent to the taiga and tundra that developed later in the Holocene.

Another peculiar grazer on the Mammoth Steppe was the saiga antelope. This gregarious creature was not a true antelope, but a steppe-adapted relative of sheep and goats. The saiga antelope was more abundant than camels, but not as common as the big three – bison, horse and mammoth.

Smallish and comical-looking with an oversized and protruding snout, saiga still thrive-on arid steppes of the Asian high plateau. Its unusually large muzzle is filled with bony membranes that recover water normally lost through breathing. This adaptation for conserving water and prewarming cold air was probably as important 20,000 years ago on the arid steppe as it is today in the saiga's semidesert environment.

Saiga pace, or rack, like camels. In this gait, the front and rear legs on one side of the body step forward in tandem, followed by the legs on the other side. This efficient type of locomotion allowed saiga to travel long distances without expending much energy. Pacing is not stable on rough terrain, however, so saiga probably did not live in Beringia's higher hills or broken country. Ideal saiga habitat would have been the flat river plains, broad valleys and gentler slopes that covered most of Alaska's treeless Interior.

If saiga antelope exploited barren plains far from water, they possibly were more abundant than their rare fossil occurrence indicates; barren environments are not favorable for preserving bones. This is an example of why paleontologists have to be careful when making biological conclusions based on fossil frequencies because the fossil record is always biased.

Not only do certain habitats preserve fossils better, certain eras in the past have more complete fossil records. Fossils from the last interstadial in Alaska, about 30,000 to 60,000 years ago, are less common, except for the end of this period when climatic and environmental conditions were in transition. During the interstadial, a temporary warming during a glacial period, trees recolonized Alaska. The acidic conditions and microbial activities of forest soils quickly destroy animal remains, significantly reducing their chances for fossilization. Beds of peat with their ancient twigs, leaves, needles and cones, preserved as thin black layers in the loess, are all that remain as organic relics of these forests.

Only a few fossils have been found from animals inhabiting Alaska during such warm periods in the Pleistocene. Teeth and bones from mastodons, large mammals distantly related to both elephants and mammoths, occasionally turn up in Alaska. Their teeth and associated plant remains reveal that mastodons ate woody browse and lived in forests or open woodlands. Enamel-covered cusps protruding from the mastodon teeth were used to pulverize woody plants. In comparison, mammoth teeth had grinding surfaces lined with numerous sharp ridges needed to cut up finer grasses. No mastodon mummies have been found in Alaska, but their trunks almost certainly did not have fingerlike tips like the woolly mammoths, since that was a grass-eating adaptation.

A giant ground sloth as big as a Volkswagen also prowled the woodlands during warm intervals. Unlike its living relative, the tree sloth of South America, these Pleistocene sloths stayed on the ground. They were sloth-slow, but they had nasty 8-inch claws that, when combined with their size, would have made any predator shy from attack. The main use of their claws may have been for foraging, but we can do little more than speculate how or what they ate. Biologists know that another species of late Pleistocene ground sloth, which lived in the American Southwest, ate stems, roots, flowers and seeds of shrubby desert plants, based on analysis of its dung found in caves. By standing on their hind feet, giant ground sloths could have reached heights greater than 15 feet, so it is probable in Alaska that they ate younger, more palatable parts of trees found in the upper branches. The sloths probably avoided competition from mastodons by browsing more selectively and diligently; their giant claws indicate that they also dug for other food, like roots.

The most bizarre Pleistocene mammal of eastern Beringia may have been the giant beaver. This gargantuan rodent weighed up to 400 pounds and was as big as a black bear. Giant beavers did not cut down trees or even build dams. Living more like a muskrat, they ate coarse swamp vegetation. They were excellent swimmers, but clumsy on land. An inhabitant of Beringia only during warmer times, giant beavers were never a true part of the steppe fauna. Instead, the swamps of Beringia were an extreme extension of its range during the last interglacial. Giant beavers were more common to the south, where they lived until 11,000 years ago. Because interglacial fossils are so rare in Alaska, no remains of giant beaver have yet been found within the state. But they likely were here, since giant beaver bones have been found just across the Alaska border along the

Like the mastodon, the giant ground sloth lived in Alaska during warm intervals of the Pleistocene. It stood on its huge hind legs to reach greener browse at the tops of trees, but its huge claws indicate that it also dug for food below ground. Details about the ecology of the giant ground sloth remain puzzling to paleontologists. (M. Long, Natural History Museum, London)

About 90 percent of the Pleistocene fossils in Alaska come from placer gold mining and river erosion. Gold miners must remove tons of silty Pleistocene sediments to reach the older gold-bearing gravels below. Because the silt overburden is frozen by permafrost and may be 150 feet thick, the miners use powerful water cannons to melt and flush it away. Entombed in this organic-rich Pleistocene muck are the remains of Pleistocene mammals that wash out from the silt. Sometimes these bones are so well-preserved that they still have chunks of flesh, hide and even hair clinging to them. Those long bones that were not scavenged and broken open may still have greasy marrow in their hollow interiors. (Mary E. Edwards)

Old Crow River in Yukon Territory.

Competition from modern, smaller beavers probably entered into the demise of giant beavers, but the disappearance of certain swamp vegetation and marsh habitat may have been more important. Beaver-chewed wood from both glacial and interglacial sediments show that modern beavers have been in Alaska for at least 150,000 years. Muskrats, porcupines and skunks also filled the forests and swamps of Pleistocene warm periods, but were not part of the ice-age steppe community.

Elk fossils have been found on the Seward Peninsula dating to the early Pleistocene, and they show up again in the middle and late Pleistocene of eastern Beringia. Their fossils are rare in Alaska, but elk would be associated with open woodlands and grasslands of the interglacials. Being one of the few deer that eat a lot of grass, elk also were able to survive on the shrubby grasslands during glacial periods. They were finally pushed out of Alaska by the Holocene's tundra and taiga, more suited to moose and caribou. Snow depth in Holocene Alaska also probably limited elk.

Along with the previous cast of mysterious leviathan beasts, a host of small mammals thrived on the Mammoth Steppe. Just like large mammals, the ecology and distribution of small mammals can tell us a lot about prehistoric environments.

Brown lemmings, collared lemmings and red-backed voles have long been inhabitants of Alaska. We can track their evolutionary stages from thousands of their teeth that paleontologists have sifted from Alaska's loess and muck. Other volelike rodents that inhabited Alaska are now extinct or live in Siberia, Europe or elsewhere in North America. Hares, too, are among the oldest Alaska mammals.

Badgers and black-footed ferrets dished out their mayhem on the Mammoth Steppe. They both are primarily grassland animals that hunt ground squirrels and other underground rodents. Arctic ground squirrels would have been their primary prey. Arctic ground squirrels now live only in high, well-drained areas north of the Yukon River and in the Alaska Range, but they once ranged widely over the Mammoth Steppe. In fact, some of the most amazing and informative fossils in all of Beringia are mummified ground squirrels preserved in their underground nests. They are still curled up like the day they died while hibernating. Grasses lining their nests help confirm the species of plants that grew on the surface. It must have been a common sight 20,000 years ago to see a ferret scurrying hole to hole in an Alaska "prairie dog town," or to see the dust and dirt flying from the hind end of an busily digging badger bent on similar gastronomic intentions.

The dominance of grazers and grassland-dependent species is a predominant feature of Alaska's late Pleistocene fossil assemblage. Today, Alaska's herbivores are primarily browsers or mixed foragers. Yet a few of Alaska's modern species found a niche on the Mammoth Steppe. We already saw how musk ox appeared there in more widespread numbers. Caribou, sheep and a primitive moose found refuge on the steppe as well.

Caribou are one of the few large mammals in Alaska today which seem to have done well in both the Pleistocene and Holocene. They have been in Alaska for a long time and may have actually evolved in Beringia during the early or middle Pleistocene. Caribou are eclectic feeders that eat a lot of tundra grasses, sedges, shrubs and lichens, and occasionally mushrooms, berries and even meat. While a steppe-dominated environment may not have been ideal caribou habitat, these cold-adapted beasts most likely took advantage of the more arctic-type vegetation that grew on Beringia's uplands or poorly drained, cool soils. We do not know if they formed the same large seasonal herds as today or if they were as migratory. Their success during the Pleistocene is clear, however, as their fossils are common from Europe, Asia, Alaska and many other states as far south as Tennessee.

Dall sheep are now limited to high alpine pastures in Alaska, but used to venture into lower elevations on the Mammoth Steppe, because of the greater grazing opportunities. While still primarily inhabitants of the uplands, Dall sheep were selective foragers of the highest-quality grasses, sedges and forbes, much like today. They were potential competitors with bison when their two ranges overlapped in the middle elevations. Sheep seldom made it down into the lowest river valleys, and bison rarely grazed on alpine ridges, but on the hills between and particularly in lush subalpine meadows the two probably crossed paths. Caribou, too, would not be far away from this upland encounter.

We used to think that moose entered Alaska from Asia early in the Wisconsinan ice age, if not longer ago. Now it appears that fossils attributed to them really belong to a more primitive and larger broad-fronted moose. Its name comes from its antlers, which had small palms attached to long beams. Modern moose have large, soft, movable muzzles adapted for aquatic feeding. The broad-fronted moose's muzzle was not as dexterous, so this moose was probably not as aquatic. It fed on browse in ravines and shrub pockets on the steppe, and it probably ate more grasses and sedges than do modern moose. Modern moose may have crossed the Bering Land Bridge about 14,000 years ago, about the same time as early humans. In fact, one theory holds both moose and humans probably entered Alaska at the same time for the same ecological reasons, the reappearance of forests.

RIGHT: *Ketchikan artist Ray Troll commemorated the woolly mammoth in a popular T-shirt design in 1988. (Ray Troll)*

BELOW RIGHT: *The mammals of the Pleistocene traveled between Alaska and Asia over the Bering Land Bridge, and in recent years Russian and American scientists have visited and exchanged information about fossil finds in their countries from this period. An American scientist shows a collection of bison, horse and mammoth fossils taken from a site on the Omolone River in northern Yakutia in the Russian Far East. This location is similar to sites in Alaska. (Paul Matheus)*

■ *Carnivores on the Mammoth Steppe*

The diversity of herbivores on the Mammoth Steppe meant greater feeding opportunities for Pleistocene carnivores. Lions stalked lone bull steppe bison in the grassy hills, while saber-toothed cats lurked in riparian brush and broken ravines waiting to ambush larger prey. On Beringia's flats, cheetahs matched their speed against wary horses whose survival depended on staying out of range of these predators. A cheetah fossil has yet to be found in Alaska, but probably cheetahs were here since they were the primary steppe predator of small, fast herbivores and inhabited the

Mammoth Steppe of northern Asia and southern North America in the late Pleistocene.

Yet one monster of a carnivore, the giant short-faced bear, would send all other Alaska steppe meat-eaters fleeing. Twice the size of a grizzly, this thug probably stole and scavenged carcasses from other carnivores. When it was finished eating, a few scraps might remain for an arctic fox or a hungry wolf. During the winter when prey populations were lowest and other predators were having tough times, the short-faced bear made the most out of every meal by cracking open and eating bones.

It is fairly clear why grazing specialists died out with the demise of grasslands, but why did a resourceful species like the short-faced bear go extinct? Why did lions and sabertooths disappear from Alaska, which has plenty of moose and caribou to hunt? The answers emerge by studying these carnivores' habits; they were overly specialized or too strict in their diet. Also, as the number of herbivore species declined in Alaska, predators had less feeding opportunities and fewer could coexist.

The saber-toothed cat Homotherium *lurked along the riparian brush and broken ravines on Alaska's Mammoth Steppe. (Z. Burian, Artia; courtesy of Paul Matheus)*

Terrestrial carnivores would be hard-pressed to survive on an all-meat diet year-round in Alaska today. The animals that become carnivore prey do not live today in large populations of evenly distributed individuals. Herbivores on the Mammoth Steppe, on the other hand, were more diverse, more evenly distributed, and, because they were less migratory, remained in a given region year-round.

A saber-toothed cat, for example, could easily kill a moose, but it would starve today waiting for one to pass by. Sabertooths, with their short, powerful hind limbs, were not built to pursue prey or sprint after them. Furthermore, their teeth were specialized for cutting meat, making it difficult for them to supplement their diets with plants. They also had small brains that were probably behaviorally wired solely for meat-eating. Caribou would be an undependable year-round source of meat because they migrate between ranges. A sabertooth could feast on caribou part of the year, but would be out of luck when the caribou moved on.

Lions and cheetahs are predators of open terrain and try to stalk undetected within range of their prey. When that minimum distance is reached, they speedily accelerate in attack, closing the gap before their prey can reach its full speed. Their success in bringing down otherwise faster herbivores depends on quicker initial acceleration. To do this, the cats need to be on firm ground. The Mammoth Steppe would have been ideal terrain to chase bison, horse or saiga, but imagine a lion or cheetah trying to accelerate and turn on cushiony tundra tussocks. Lions and cheetahs suffer the same strict meat-eating specialization as sabertooths, so making up for lean times with blueberries and green grasses was not an option.

Filling out a diet with berries and grass sounds like a strategy that perhaps the short-faced bear could have used. But work at the University of Alaska analyzing chemical composition of short-faced bear bones has shown that they, too, were strict carnivores. We speculate that if these bears could have hibernated through the lean winters, they may have survived to this day.

Short-faced bears were amazing creatures. Part of a subfamily of bears native to the New World, they dominated North America for almost a million years. They were not built like modern bears. Instead, they were lean and long-legged, with huge heads and short, powerful jaws. Some paleontologists have suggested these bears looked and behaved more like cats. But considering the abundance of other more proficient predators, short-faced bears may have been scavenging specialists who used their huge size to dominate carcasses and chase off other large carnivores.

Short-faced bears did not have to contend with brown bears for most of their history, because brown bears crossed the Bering Land Bridge from Asia fairly recently, perhaps 60,000 to 80,000 years ago. Brown bears were one of the animals that migraterd from Asia to Alaska and became trapped here until the North American ice sheets receded. For this reason, they did not reach the rest of the United States or Canada until the Holocene. The purely meat-eating short-faced bear would have been completely dominant over brown bears. Meanwhile, the opportunistic brown bear would have done what it does best: eat vegetation, small mammals and fish. By hibernating, it significantly reduced its annual calorie needs, giving it an even greater survival advantage. This ecological plasticity is why brown bears survived Pleistocene extinction and still thrive in Alaska.

Black bears are forest-dependent and did

not live on the Pleistocene steppe. However, they may have been in Alaska during previous interglacials. There are no fossils to support that contention, but they have a long fossil history in the rest of the United States and southern Canada.

Two small carnivores, the lynx and wolverine, hunted on the Mammoth Steppe and still live in Alaska. Lynx took advantage of available snowshoe hares, also a common mammal of Pleistocene Alaska. In fact, earlier in the Pleistocene, snowshoe hares were twice their current size, and it would have been exciting to watch a lynx chase one through a willow thicket. Like today, lynx populations probably fluctuated in response to highs and lows in hare numbers. The lynx is the only surviving Pleistocene carnivore that is purely a meat-eater, but it survives because it specializes on smaller, more abundant prey.

Wolverines prefer to eat meat, but can live in a variety of habitats as they are opportunistic omnivores. Living in forest and tundra today, their populations never reach high densities. The same was probably true when they lived on the Mammoth Steppe. Wolverines can be ferocious predators and can even kill vulnerable caribou, which are many times their size. Wolverines also use their powerful jaws to crunch bones and, like the short-faced bear, probably took advantage of the mammal carcasses strewn about the Mammoth Steppe.

Lastly, we consider the wolf, Alaska's long-term, resilient carnivore. For maybe as long as a million years, wolves have watched the parade of mammals come and go through Alaska. Whether transient species en route between Asia and North America, short-lived evolutionary experiments, or permanent northern inhabitants, flesh from almost all of them fed wolves at one time or another.

Able to pursue and kill prey over twice their size, wolves can easily switch to diets of tiny rodents, insects and berries. Social and intelligent, wolves are scrappers and highly adaptable. Wolf packs form or disband to suit the situation. At times when the Mammoth Steppe was filled with other large carnivores, wolves may have been solitary and opportunistic. On the other hand, wolves and lions were the only social carnivores on the steppe, and the wolves' ability to form large packs may have been their secret to success. Despite being smaller than other carnivores, a large group of wolves could have challenged the fiercest solitary carnivore for its prey.

Furthermore, no other carnivore in Beringia was designed for the long-distance pursuit strategy which wolves used to tire their prey. The lightly built, thin-legged wolf is a cursorial master who could follow and harass for days animals like camels, broad-fronted moose, caribou and the occasional solitary bison or musk ox. When the victim finally succumbed to hunger and stress, the wolf could finish it off.

We end with a true scene of a modern wolf on Alaska's North Slope. She feeds and harbors her pups in the same den used by her Pleistocene ancestors. One pup sits on an earthen mound outside the den chewing on the leg bone of a caribou calf. The mound has formed during many millennia by the activities of countless previous pups. A few inches beneath the surface rest bones of horses, camels, musk oxen, caribou, bison and woolly mammoths chewed by her Pleistocene forerunners on the same, yet distant, land. ■

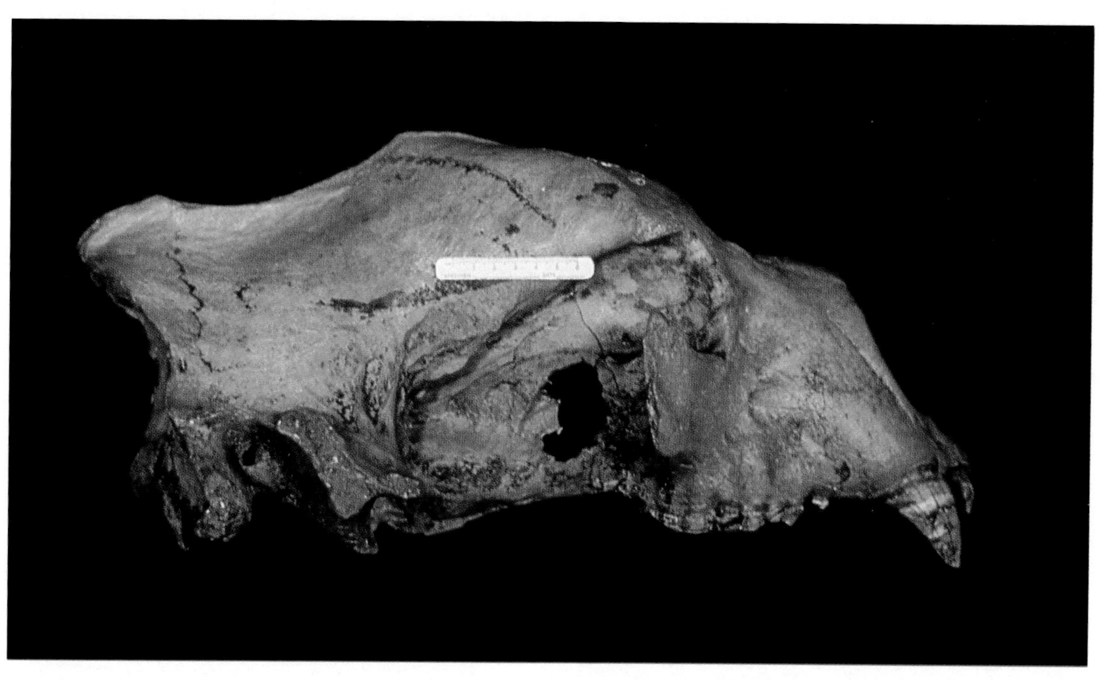

This skull is from the original Alaska bear — the giant short-faced bear, which may have lived here for almost a million years. This huge, but lightly built bear was twice the size of a grizzly and purely carnivorous. Its long legs and light build made it an efficient traveler, and it probably survived by roaming vast territories and using its super-sensitive nose to search for animal carcasses. Letting other carnivores do the killing, the giant short-faced bear may have evolved its huge size to help it steal and defend carcasses from other carnivores. (Paul Matheus)

The First of the Last:
Pioneer Human Settlers on the Last Frontier

By Dr. William Workman

Editor's note: *Dr. Workman is a professor of anthropology at the University of Alaska Anchorage. Karen Wood Workman contributed substantially to the preparation of this account.*

The hunters who crossed from Siberia to Alaska toward the close of the Pleistocene were both the first and last of their kind. They were the first humans to enter North America and they were the last who would see an entire continent where none of their kind had ever lived before. Some of their descendants would traverse both North and South America, covering 16,000 miles with surprising speed.

For more than 2 million years tool-making humans have lived in subtropical Africa. By 1 million years ago some had left Africa to colonize Asia and, perhaps somewhat later, the peninsula Europe. For the next million years humans gradually moved north across Eurasia. By 40,000 years ago pioneer settlers made their way to Australia, leaving the Americas as the only continents yet to be peopled. A number of times in the past million years the northern and temperate zones of the Earth endured great cold waves, each of which lasted the better part of 100,000 years. Continental glaciers, great masses of ice thick enough to bury mountain ranges, expanded during these times, driving people back toward the south. Gradually some people learned to cope with the seasonal cold of winter, and eventually with the enduring cold of the northern parts of the Earth. By 25,000 to 30,000 years ago, Eurasians were hunting in bands on the treeless northern grasslands that stretched from the Atlantic to the Pacific, although the maximum cold of the last glaciation forced a final southward retreat between 18,000 and 24,000 years ago.

While the grasslands of the ice-age landscape abounded with bison, horses, mammoths and other mammals, the entrance requirements were high for human hunters attempting to live there. Minimum requirements, in addition to big-game hunting skills, included the use of fire, the ability to construct or locate warm shelters, and warm clothing. Hands and feet are particularly subject to cold damage, so boots and mittens were a prerequisite for northern life. Since edible plants are neither diverse nor abundant in northern regions, the pioneers needed to be master hunters and, perhaps, fishermen. No metals were yet used, so the weapon technology depended on killing points of stone, bone, antler or ivory attached to wooden shafts. Since these hunters had to move frequently and since suitable stone is locked in frozen ground during long northern winters, it is not surprising that early northern technologies were sparing in their use of stone. Microblades and cores were a common, but not a universal, example of such technical efficiency. Many small, sharp-edged, parallel-sided flakes (microblades) were struck from carefully prepared, small stone blocks (microblade cores). Microblades or microblade segments could then be inserted into wood, bone or antler handles and weapon heads and used for many purposes. Much attention focuses on the stone spear heads, knives and other weapon parts, but perhaps most important in any stone technology are the special tools to prepare skin, wood, bone and other perishable substances.

Entry into North America was delayed because the entrance requirements were so stringent. By far the most reasonable port of entry from Asia to America is the Bering Strait area well north of 60 degrees. During ice ages, lowered sea level created the Bering Land Bridge, and opened a vast plain called Beringia. At these times Alaska and northwestern Canada were an eastern peninsula of Asia, separated from the rest of the Americas by ice farther south.

By any reasonable calculation the first Americans appeared late in human history. We do not know how often human groups migrated across Beringia or how many groups were involved. One prominent theory, based on the physical characteristics and broad language groupings of modern Native Americans, suggests three migrations. Last to come were ancestors of the Eskimos and Aleuts, earlier were ancestors of the Athabaskan, Tlingit and some other northwestern Indians, and, earliest, the ancestors of all other Native Americans. Alternatively, other linguists visualize many migrations spaced throughout many thousands of years. There are other possible reconstructions as well.

Our interest in the pioneer settlers of North America far outstrips our knowledge of their ways of life. We know that they came when Alaska was different from the Alaska we know, but exactly what environments they confronted depends on when they came. Although any reasonable scenario for peopling North and South America must place Alaska as the point of entry, we have yet to find sites here significantly older than sites in the southern American plains. Scholars disagree about exactly when humans arrived in the New

ABOVE: *Archaeologists' tents cluster around an excavation site at Reese Bay on Unalaska Island in the Aleutians. Scientists working here estimate that humans have occupied this area for about 2,000 years. (Douglas W. Veltre)*

LEFT: *Most anthropologist and archaeologists in Alaska focus their attention on recent prehistory because more artifacts are available to guide their reconstruction of these lifestyles. Among these clues are artifacts excavated from the Korovinski site on Atka Island in the Aleutians: from left, two chipped stone knife blades, bird bone awl, two chipped stone adzes or scrapers, one chipped stone knife blade, and at bottom, three bird bone sewing needles. (Douglas W. Veltre)*

World, a sure sign that the available information is too fragmentary to be compelling. We know that elephant hunters occupied the eastern flanks of the Rocky Mountains by 11,500 years ago and that humans had reached the tip of South America by 11,000 years ago. We think they must have come to Alaska by at least 12,000 years ago. Peopling of continents is not a race. There would have been no inducement to see who could be the first to arrive in Patagonia. Few scholars seriously suggest a human presence in the Americas before 50,000 years ago, while a highly conservative position would bring the first settlers across Beringia only a few centuries before 11,500 years ago. More workers probably tend to the conservative (late entry) rather than the liberal view, although few of us are altogether happy about this situation.

The first Alaskans entered a world that no longer exists. Sea levels were many feet below today's, with attendant rearrangement of the geography of the Alaska coast. The vast black spruce-muskeg plant community that dominates much of the Alaska interior had not yet developed, although dwarf willow and dense thickets of other woody plants may have been widely distributed. Bison and elk were encountered in addition to caribou and mountain sheep. The status of moose is less clear. It is a matter of debate whether or not man shared the landscape with the woolly mammoth, herds of small shaggy horses and other prominent but doomed Pleistocene animals. In places the landscape was recently freed of glacial ice and would have been raw, dusty and windswept, although substantial areas of interior Alaska were not glaciated in the late Pleistocene. Summers, while short, would have been warm. Although we usually think of the first Americans as hunters of terrestrial game, it would have been possible for sea mammal hunters to migrate from Asia to America along the southern coast of

BELOW LEFT: *Richard Nelson and Mazakazu Yoshizaki excavate the 8,000-year-old core and blade site of Anangula in 1963. Anangula, on an island near Nikolski, is the oldest-known site of human habitation in the Aleutians. The term "core and blade" refers to a distinctive way of making stone tools that involves removing long, narrow flakes from a specially prepared piece of stone. (Allen McCartney)*

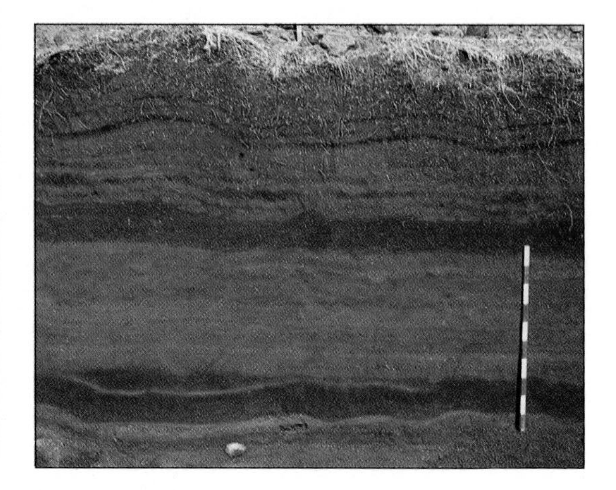

BELOW: *This meter stick rests on the cultural horizon at the Anangula core and blade site that is buried beneath 6 feet of subsequent ash layers; the dark brown ash directly above the cultural horizon represents a major volcanic eruption in the region that caused the abandonment of the site. (Allen McCartney)*

ABOVE: *Prehistoric people built this 12-foot-high hunting blind on the rim of Twin Calderas on the Seward Peninsula to control the movement of animals they were hunting. (Jeanne Schaaf, National Park Service)*

RIGHT: *As early as 9,500 years ago, hunters waited for passing prey in the limestone caves above Trail Creek in the Bering Land Bridge National Preserve on the Seward Peninsula. About 4,500 years later, members of a different culture also took shelter in the caves, whose history of use extends into modern Inupiat Eskimo times. (Jeanne Schaaf, National Park Service)*

Beringia. It is far from certain that late ice-age humans anywhere in the world had developed either an interest in or the skills for open-sea hunting and fishing however. Even if humans came before the glaciers began their retreat about 14,000 years ago, substantial coastal areas would have been free of glacial ice. If the pioneers had sea-hunting capabilities and good boats, they might have proceeded south along the partially deglaciated coast. Whether they actually did this remains a hotly debated question.

The evidence we have about the way of life of the pioneer settlers suggests small, highly mobile groups. Although some Siberian ice-age hunters constructed solid semisubterranean huts, we have yet to find any such durable shelters in early Alaska. Much speculation, based on little hard evidence, has been lavished on the subsistence economy of the first Alaskans. Unusual is the site from these remote times that yields the bones of animals in addition to stone tools. In the absence of other reasonable options we can be sure they were sophisticated hunters, but the degree to which they depended on now-extinct animals is unknown. Rare sites with preserved bone food debris, like Broken Mammoth, indicate that a diverse spectrum of small mammals, birds and fish were exploited in addition to big-game animals in late ice-age times.

Little evidence of human activity has been recognized as yet in interior Alaska in early postglacial times, i.e. between 7,000 and 10,000 years ago. By around 6,000 years ago a new technology, probably derived in part from the south and east, was established along the Kobuk River, in the Brooks Range and eventually widely throughout the forested interior and uplands. A hallmark of this Northern Archaic technology is side-notched stone weapon heads, probably spear points. Some suggest that the Northern Archaic is implicated in the origins of northern Athabaskan Indian cultures that become fully visible only in the last 1,000 years.

Evidence of a human presence in the archipelagos of southeastern Alaska before 10,000 years ago suggests good boats and the craft of the sea hunter and fisherman. The Aleutian evidence reviewed by Douglas W. Veltre indicates this subsistence pattern was present at Anangula, the oldest known archaeological site in the Aleutians, between 7,000 and 8,000 years ago, and was in place widely in the Aleutians between 3,000 and 4,000 years ago.

LEFT: *This area has been plotted for excavation at the Mesa site. Prehistoric hunters, sitting around a fire and crafting their weapons, could look out across stream-laced lowlands to neighboring Brooks Range foothills from their lofty perch. In the lowlands grazed the animals that became the hunters' prey. (Sherie Dye)*

LOWER LEFT: *The oldest known site of human habitation on the southern Kenai Peninsula, this tiny, 5,000-year-old encampment on the shores of Kachemak Bay is just 3-plus feet deep. For unknown reasons, soil never accumulated atop it. (Janet R. Klein)*

LOWER RIGHT: *St. Lawrence Island has yielded abundant clues about the activities of prehistoric people. This ancient gravesite near Gambell, marked by whale bones, is from the Punuk period, about 900 A.D., and was excavated by Professor Hans Georg Bandi of the University of Bern, Switzerland, in 1972. (Chlaus Lotscher)*

Unquestioned Pacific maritime hunters of the Ocean Bay cultural tradition were established along the base of the Alaska Peninsula and on the Kodiak archipelago by 6,000 years ago. On Kodiak this way of life, which spread to outer Cook Inlet and to Lake Iliamna by 4,500 years ago, endured until about 4,000 years ago and contributed to the formation of the later prehistoric cultures in this area. The origins of the Ocean Bay tradition are unclear. Some areas, such as the Cook Inlet basin, appear to have been repeatedly colonized from adjacent areas but were not able to support continuous cultural development for thousands of years. Other areas such as Prince William Sound appear to have been settled only within the last several thousand years.

The first inhabitants of the tundra and frozen coasts surrounding the Arctic Ocean from Alaska to northern Greenland appear abruptly about 4,000 years ago. Their tiny, jewellike implements, giving rise to the name the Arctic Small Tool tradition or ASTt, suggest to some that they were recent immigrants from northeast Asia. The ASTt appears to be implicated in the ancestry of Eskimo-speaking peoples although the history of ancestral speakers of Eskimo-Aleut languages appears to be too complex and varied to be accounted for through one migration from Asia. It is probable that regional varieties of Eskimo culture developed in North America with technological contributions from the ancient sea-hunting cultures of the northeastern Pacific Ocean and possibly later northeast Asian cultures as well as the ASTt.

In summary, we have much to learn about the first Alaskans, among whom must also be numbered the first Americans. Although the small, deeply buried sites that have survived can tell us much about their age, technology, survival techniques, group size, connections with other groups and the environment in which they lived, other interesting questions are more difficult to answer because no literate human ever witnessed the infiltration of an entirely new continent by human settlers. What were the psychological dimensions of being the first humans to enter a river system or bay? How prone were these people to strike out purposefully into the unknown? How similar to modern game animals was the behavior of animals who had never encountered a human hunter before? To these relatively unique questions can be added others with which the materialist approach of archaeology has difficulty dealing. What languages did these early Alaskans speak, and what tales did they tell around their camp fires? How did they view their position in the natural world and the world of the spirits? What conclusions had they drawn about the realities of birth, death and the inexorable passage of time and with what actions did they express their beliefs?

Finally, it must be emphasized that the question of early man is only one of a number of interesting and important topics in Alaska archaeology. The history of use and occupancy of this land in recent prehistory is both more approachable (because the information is more abundant) and perhaps of greater relevance to the Native Alaskans who live here today. Most Alaska archaeologists deal only occasionally with the earliest occupants, concentrating most of their efforts on an attempt to understand the more recent past. This task, while difficult, carries its own rich and enduring rewards. ■

BELOW: *Archaeologists gather near the junction of the Gulkana and Copper rivers to excavate an early Athabaskan site. (Steve McCutcheon)*

RIGHT: *Archaeologists didn't have to dig far to find evidence of a prehistoric tent site near Galbraith Lake in the Brooks Range. (Steve McCutcheon)*

Clues from an Ancient Mesa

By Lee Dye
Photos by Sherie Dye

Editor's note: *Lee Dye is a free-lance writer, and a former science writer for the* Los Angeles Times. *He and his wife, Sherie, a photographer, split their time between Juneau and the American Southwest. This article was written in fall 1993, and updated after the 1994 season at Mesa. Also in 1994 the Bureau of Land Management withdrew 2 square miles of undeveloped public lands around the Mesa for 20 years to protect its archaeological, historical and cultural resource integrity.*

Michael L. Kunz had spent the day slogging across the tundra 150 miles north of the Arctic Circle when he first came to the place that would change both his life and our perception of what the earliest Americans were like. As a young archaeologist with the U.S. Bureau of Land Management, Kunz was leading an expedition near Ivotuk, a desolate outpost 200 miles south of Point Barrow. The year was 1978, but it would be more than a decade before he would realize the significance of what he found when he hauled his tired body up the steep slopes of an austere rock outcropping now known as the Mesa.

The Bureau of Land Management has jurisdiction over the vast, oil-rich regions of Alaska northlands, and federal regulations required that archaeological surveys be conducted before drilling could be carried out in the pristine areas of the Arctic. A few sites of archaeological significance had already been found, part of the fortuitous fallout from the oil boom, but Kunz and a colleague, Dale Slaughter, had struck out on that cold summer day in 1978.

Their helicopter had dropped them off early that morning, and they had been fighting their way across the tundra, stepping among small tussocks that can collapse under the weight of a human foot and twist an ankle. Both men knew the only way to do their job was to walk the land, and so they had done that for hours.

"It had been a long day and it was getting late, and we had come to this place," Kunz says, pointing toward the imposing rock that juts up out of the tundra like a natural cathedral. "We hadn't found much, and we were both very tired. I told Dale I'm going to climb up there, and he says 'I'm not climbing up that damn thing'."

The two men had been concentrating on hilltops where prehistoric hunters would most likely have left traces of their presence. Exhausted by the long day, Slaughter sat down at the base of the outcropping, but Kunz wasn't ready to quit.

"I went up, and right away I saw there were flakes (of stone) on the surface," he says. "They are the result of tool manufacturing, so I knew there was a site there. As I glanced around, I could see that there was quite a bit of stuff there.

"I'm walking along and all of a sudden, bingo, there's a projectile point lying there. I picked it up and I thought, that looks kind of like Paleoindian stuff." Kunz had studied archaeology at Eastern New Mexico University,

ABOVE: *Bureau of Land Management archaeologist Michael Kunz stands at the site he discovered in 1978. He carries a shotgun because of the presence of bears at the site.*

ABOVE RIGHT: *Archaeologists camp beneath the Mesa, on top of which were excavated some of the oldest artifacts yet to be found in Alaska.*

so he knew something about the tools used 11,000 years ago at the oldest sites in the Southwest occupied by early Americans, known as Paleoindians. Other artifacts found in the Arctic had been so different from those found in the Southwest that archaeologists believed the first Americans developed their distinctive tool-making methods after they traveled south and into the heartland of the United States.

"When I came to Alaska in 1970, I figured I had seen the last of Paleoindian," Kunz says.

Kunz found a total of six spear points lying on the surface of the Mesa, and although he was excited about the discovery, he initially doubted that the points would prove significant because it was impossible to date them. To determine how old such an artifact is, it must be associated with organic materials that can be dated, and this is difficult in the Arctic because the soil is so thin and subject to such a harsh climate that materials containing datable carbon are rarely found with the artifacts. The spear points found by Kunz could have been dropped there 12,000 years ago, or yesterday.

But he was haunted by the discovery, and returned the following year with a small crew. They discovered three hearths and 11 projectile points buried in or near the charcoal left by the fires. The fires must have been created by humans, because the region is treeless and wildfire is unknown atop such land forms. Prehistoric hunters must have carried the firewood up the hill from willow-lined stream banks below. The charcoal could be dated using carbon 14 techniques, but in those days the process took so much charcoal that the hearths had to be combined to provide enough material. The date came back from the laboratory at 7,620 years. That was interesting because little is known about that time period, but the date was not old enough to be of Paleoindian age.

In 1980, Richard E. Reanier, a doctoral

student in anthropology at the University of Washington, joined Kunz. The two conducted additional tests at the site. After that season,

LEFT: *Francis P. McManamon, archaeologist with the U.S. Department of the Interior, sifts for artifacts. The Mesa site has proved prolific, yielding 2,000 to 3,000 waste flakes and stone tools.*

BELOW: *Focus of much attention among archaeologists in recent years, this mesa on the north side of the Brooks Range and near the western edge of Gates of the Arctic National Park has been the site of hunting camps for prehistoric people off and on for at least about 11,000 years.*

"we drifted away from (the Mesa)," Kunz says.

Nine years later they came back and found an additional hearth. In the meantime, a new dating technique had emerged that counts the carbon 14 atoms. Since carbon 14 decays at a uniform rate, the amount of carbon 14 can tell precisely when the organic material was alive, and hence when the early hunters had actually collected their firewood. The improved dating technique, which requires a much smaller sample, produced a surprising figure. The carbon in one of the hearths came from a fire 9,970 years ago.

In the following years additional hearths yielded even older dates, and in 1992 the two archaeologists hit the jackpot. One of the hearths was dated at nearly 11,700 years, indicating it was among the oldest sites yet discovered in North America, even older than those in the Southwest and the Great Plains. It was also the most productive hearth found, yielding 2,000 to 3,000 waste flakes and stone tools including several spear points. Some of the points had distinctive fractures caused when rock is exposed to heat, further cementing the relationship between the datable hearth and the artifacts.

"It can't get any better than that," Kunz says. "We got stuff right in the charcoal itself, and it got thrown in there when the fire was burning."

Meanwhile, other well-documented sites had been discovered in Alaska that also yielded ancient artifacts. The link between the Arctic and the heartland of America had taken a major turn.

"All of a sudden, whoa, we've got full blown, recognizable Paleoindian stuff in Alaska between 10,000 and 12,000 years ago," Kunz says. "That's just a totally new concept."

It suggests, he adds, that the first Americans who crossed the Bering Land Bridge were not of a homogeneous culture. They more likely represented many cultures and came from many areas before they crossed Siberia and, unknowingly, entered the New World.

The discovery of the Mesa site does not answer one of the most daunting questions in American archaeology: When did the First Americans arrive here? Some believe the migration took thousands of years, possibly beginning 50,000 years ago. That early migration argument is supported by archaeological sites in Pennsylvania, the Yukon and South America that range from 16,000 to 40,000 years old. But many archaeologists doubt the reliability of those early dates. What is most troubling is the fact that the archaeological record does not show the proliferation of sites more than 11,500 years old that one would expect to find if the migration began tens of thousands of years ago.

The debate has been fueled partly by other discoveries in Alaska by John F. Hoffecker of the Argonne National Laboratory in Illinois, and Roger Powers and Ted Goebel of the University of Alaska Fairbanks. They discovered Paleoindian-age artifacts at several 11,000-year-old campsites north of Denali National Park. In a 1993 report in the journal *Science*, the three anthropologists side with archaeologists who "have steadfastly maintained that compelling evidence for sites older than 11,500 years has yet to be found."

They argue that archaeological research throughout the years should have turned up a pattern of older sites if they do in fact exist. That negative evidence -- the absence of reliably dated older sites -- indicates that there were no humans in North America prior to about 12,000 years ago, they contend. The Mesa find will not resolve that argument, and it will probably go on for many years.

The world of archaeology is littered with the bodies of experts who spoke too quickly, and while Kunz and Reanier were confident of their findings, they wanted additional support before formally presenting their work.

"I said we've got to take this show on the road," Kunz recalled. The two set out across the country, showing their artifacts to experts in various fields. They called on C. Vance Haynes, a carbon-dating expert at the University of Arizona; George C. Frison, a hunting guide turned archaeologist at the University of Wyoming and one of the country's top experts on early hunting techniques; Phillip H. Shelley of Eastern New Mexico University, an expert on early tools; and others. In separate interviews, all said they believe the site is more likely only about 10,000 years old instead of 11,700, but all confirmed that they believe the artifacts are truly Paleoindian and the site is very significant.

"We came back with a major double red alert," Kunz says.

Archaeology consultant Richard E. Reanier shows George C. Frison of the department of anthropology, University of Wyoming, carbon from one of the hearths at Mesa.

Kunz needed funding to continue working the site. "I had been begging and borrowing to get up here," and his superiors at the BLM saw an opportunity to improve the image of the nation's largest landlord. So the agency's top officials in Washington called a press conference.

"BLM wanted us to wave the flag," Kunz says.

Kunz is an affable, confident man of 52 who seems to enjoy dealing with the public,

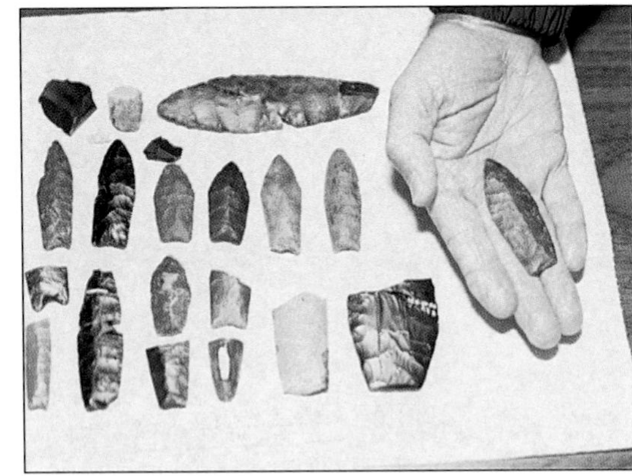

LEFT: *Daniel Mann, geochronologist at the Quaternary Research Center, University of Alaska Fairbanks, holds an ice core sample taken from soil around the Mesa.*

ABOVE: *This collection of projectile points is among the thousands found at the Mesa site.*

but nothing could have prepared him for the press conference, not even the tumultuous years he spent as the bureau's chief environmental field monitor during oil and gas exploration in the National Petroleum Reserve-Alaska. He had expected half a dozen or so reporters to show up, but when he arrived he found an auditorium jammed with television cameras and news hounds. Archaeology is a hot topic these days, and when a federal agency calls a press conference to announce a major finding, everybody shows up.

"The story appeared everywhere, in 2,500 newspapers in the United States and Canada," Kunz says. The press conference made the Mesa site the most celebrated, if not the most important, archaeological discovery in Alaska. That notoriety has caused some bitter feelings.

"They hyped it," says Brian Fagan of the University of California, Santa Barbara, who believes the Mesa site is significant but no more so than other Alaska discoveries at a group of sites called the Nenana Complex. All suggest that some humans arrived in the New World at least 12,000 years ago.

The press conference did achieve one of its fundamental goals. The bureau gave Kunz enough funds to return to the Mesa in summer 1993 and bring the experts from the Lower 48 with him. On a cold summer day, they gathered at the Mesa to render up their judgement.

"It's obviously a significant site," says Vance Haynes, peering out from under a safari hat and brushing aside a swarm of mosquitoes. But Haynes is troubled by the fact that all the artifacts are so similar, suggesting they were made by only one or two artisans. If the Mesa had been occupied for 2,000 years, as the carbon dating indicates, their technology should have evolved somewhat, but the artifacts from the oldest hearths look almost exactly like those from the youngest. That has convinced Haynes that the site was probably occupied for only a few seasons, and most likely around 10,000 years ago at the end of the Pleistocene. At about that time the melting glaciers had caused the seas to rise enough to cover the Bering Land Bridge, so if the Mesa's hunters did in fact cross the bridge into North America they may have been among the last to do so.

Most likely, it was a band of just three or four families, led by a shaman, Haynes says. However, he was probably not a chief in the popular sense of the word, and held only loose authority over his followers. Those who sat on top of the Mesa and made the tools so critical to their survival probably had no higher rank than those who had mastered the art of stitching together animal hides to protect them from the cold.

George Frison, a mountain of a man who spent half his life as a Wyoming rancher before turning to archaeology, talks of how little we know of those early hunters as he fondles one of the spear points found in a hearth. Archaeology is advanced in inches, not miles, he notes, and the Mesa is only one part of the puzzle. But the size of the spear point tells him something.

"These were hunters," he says with the ring of authority that comes from having been a hunter himself. "And these points were made for large animals."

Those early travelers picked the Mesa, he adds, because it towers over the gentle, sloping hills in the northern shadow of the Brooks Range. From the top, they could see in every direction, watching for the great bison or other large animals as they made their way along the streams that twist through the valleys below.

They built fires to keep themselves warm, and for a few seasons at least, they sat by the fire and worked on their weapons. Chipping away at the chert they found along the riverbanks, they shaped the spear points that would give them the power to bring down the great beasts that fed their families.

Somewhere below, possibly along a stream whose rushing water has long since wiped out the remains of their settlement, others stitched together the animal hides that protected them from the cold. In time, they moved on, most likely because the animals they hunted had also traveled on, and they abandoned the Mesa and never came back.

The wind slowly filled their hearths with fine dust, and the delicate flowers that paint the tundra in pastel colors each spring knitted their roots together, entombing the place where they had sought warmth and watched for game.

As they left, some of their spear points were dropped on the ground, and they remained there for at least 10,000 years until a young archaeologist decided to climb just one more hill.

Kunz and Reanier will return in 1995, and possibly one or two years after that, until about 10 percent of the Mesa's top has been excavated. Then he will leave the site for other generations, armed with technologies we now know nothing about, to come again and search for clues about those who came this way so long ago. ■

BELOW: *Russian archaeologist Sergei Slobodin uses a trowel and dustpan to uncover more artifacts at Mesa.*

BELOW RIGHT: *Artifacts prompt discussion among (from left) George C. Frison, Michael Kunz, and C. Vance Haynes Jr.*

Seward Peninsula Prehistoric Lifeways

By Jeanne Schaaf

Editor's note: *Jeanne Schaaf is an research archaeologist with the National Park Service.*

During a series of ice ages throughout the past 2 million years, low sea levels intermittently exposed a tundra plain, a bridge linking Asia and North America. This land bridge at its maximum extent spanned more than 600 miles from the Aleutian Islands to the Arctic Ocean. Today's Chukotka and Seward peninsulas and the Bering Sea islands from the Pribilofs to the Diomedes were areas of high relief and scattered small peaks on the otherwise low plain.

Ranges of Old and New World flora and fauna extended eastward and westward across the bridge. A detailed snapshot of this environment during the last glacial period is provided by the exciting discovery of a 17,000-year-old vegetated surface buried under volcanic ash on the Seward Peninsula in what is now the Bering Land Bridge National Preserve. Pollen, plant debris, insect and animal fossils recovered from this buried landscape reveal a dry or steppe-tundra environment with grasses, sedges, sage, legumes, willow and alder.

Small groups of hunter-gatherers probably were living on the land bridge by 15,000 years ago, and perhaps much earlier. They and other predators such as wolves, bears and lions hunted the woolly mammoth, bison, caribou, antelope, horse and other animals foraging on the rich plain. Caves in the limestone formation at Trail Creek on the Seward Peninsula contain hunting implements and bones of prey left by hunters as early as 9,500 years ago, shortly after the land bridge was last flooded. The caves were not used as long-term residences, but were used as temporary shelters by hunters on the move. Their bone and antler points, slotted for microblade insets, belong to an Asian/Siberian and Alaska tool technology called the American Paleoarctic tradition. Hunters later used the same caves for shelter from 5,000 years ago through historic Inupiat Eskimo times. Trail Creek Caves is the oldest known archaeological site on the Seward Peninsula and also provides an unparalleled 15,000-year-old paleoecological and paleontological record. Serpentine Hot Springs in northcentral Seward Peninsula is the location of small prehistoric camps with stone tools that may also fall within this tradition.

There is a gap in the Seward Peninsula archaeological record after the American Paleoarctic tradition until about 5,000 years ago. Elsewhere in Alaska people of the Northern Archaic tradition were hunting caribou and fishing, perhaps with nets, by 6,500 years ago. Although sediments at Trail Creek Caves date to this time period, no associated artifacts have been found, suggesting that the Seward Peninsula was not occupied by Northern Archaic people.

The makers of finely chipped stone tools, whose seasonal round included coastal sea mammal hunting, inland fishing and caribou hunting, appeared as early as 5,000 years ago at Kuzitrin Lake. This is the earliest date in Alaska for the Arctic Small Tool tradition, which spans 4,000 years and includes several different cultures, including, from the earliest: the Denbigh Flint Complex, Choris, Norton and Ipiutak. Seasonal sea-mammal hunting camps and residential sites representing these groups are found along the Seward Peninsula coast from Safety Sound to Deering. The beachridge complex at Cape Espenberg preserves a long record of culture history from Early Arctic Tool tradition sites through modern Inupiat Eskimo sites. These ridges are also an important proxy record of climate change because their geomorphology preserves a record of changing prevailing storm and wind patterns.

The northern portion of the peninsula contains a broad plateau of weathered basaltic lava flows ranging in age from more than 5 million years to as recent as 2,000 years. Five large maar, or rimless, crater lakes, the largest known in the world, occur within this area. Gently rolling hills of low relief are punctuated by broad-domed summits of ancient volcanic cones. This landscape is dotted with hundreds of stone structures constructed by caribou hunters during the past few thousand years. These structures, which include hunting blinds or "inuksuit," cairns and caches are found on the volcanic cinder cones and boulder fields of weathered lava flows. Some are as simple as loosely piled slabs and others are spectacular monuments more than 12 feet high that tower above the landscape from the rims of ancient

From right, John Morehead, director, Alaska Regional Office, National Park Service; Ken Adkisson, former supervisor, Bering Land Bridge National Preserve; and Ferdie "Fritz" Wohlwend explore Cave 9 in the limestone rock above Trail Creek, perhaps the most well-known archaeological site on the Seward Peninsula. At left a screen and a bit of scree from Helge Larsen's work at this cave in the 1930s is visible. (Jeanne Schaaf, National Park Service)

calderas. Substantial interior villages with as many as 25 houses are associated with these stone structures.

Although research and analysis of the prehistoric sites across this landscape are far from complete, it is thought that many of the stone structures were components of large communal caribou hunts. It is clear that some of the structures were built in late prehistoric times, 100 to 400 years ago, yet it is also possible that some represent similar activities conducted

Layered tephra covers lighter-colored paleosol, the 17,000-year-old land surface, at Swan Lake within Bering Land Bridge National Preserve. (Claudi Hoefle, courtesy of Jeanne Schaaf, National Park Service)

The Way of the Hunter

By Herbert O. Anungazuk

Eskimo James Kivetoruk Moses used pencil, pen and ink to depict an Eskimo's encounter with a polar bear on the sea ice. (Anchorage Museum, Photo No. 72.117.1)

Editor's note: *Herbert Anungazuk is a Native liaison and heritage specialist with the National Park Service in Anchorage. This is his account of what a hunting trip might have been like for a prehistoric hunter in western Alaska.*

The hunter walked stooped down the long tunnel of his snug earthen house. He could hear the ferocity of the blizzard, but he knew as he neared the exit that the storm was weakening. From white-out to visible skies amid windblown clouds, the last storm of winter was expending itself. Visibility was poor, but the horizon was dark, indicating that the sea far from shore was free of ice. He glanced north for telltale signs that only a hunter can read. Whitewashed skies told that the ice pack had returned, but he could see a dark sliver on the horizon. The dark sliver, an open lead in the ice, meant that an avenue was open to the sea mammals that would pass through these waters soon on their yearly migration to feeding grounds to the north. And, with the spring hunts so near, the hunter again waited the arrival of the great whale.

For days the storm had raged, but the hunter knew it was time to prepare. He returned to his house and changed into clothing he had worn all winter in his quest for seal. Swiftly he gathered the barest necessities to scout the ice where the alliance between man and whale would be renewed. The white bird (snow bunting) had been seen before the storm, and from his grandfather he knew that the first sighting foretold the coming of spring and the whale.

The water was distant across the shore ice, but eventually he neared the lead. The hunter tested each step with his staff, as the drifted snow did not reveal the dangers of thin ice. Hunters had been known to plunge through the ice and be lost. The wind was constant, but within its sound he could hear the breath of many whales. He stepped slowly toward the water, even though he knew that the sound of his footsteps carried loudly over ice and snow. As the hunter inched slowly over the pressure ridge, he did not expect that another hunter was also curious. At the same moment, man and white bear peered over the ice and saw many white backs appearing and disappearing in the cold, frothy water. The elder brother of the great whale had arrived — the white whale always reaches the hunter's waters first.

Few hunters have seen the white bear hunt its prey. The hunter had never seen, but had heard the old ones tell stories about the hunting prowess of the bear. He realized that the rarest of opportunities had come to him to learn the skills of another hunter so he watched as the great white bear inched slowly toward the water. The hunter noted that the bear had covered its nose with his massive paw. To be seen would be untimely and would result in the failure to fill an empty belly. The hunter did not see the bear spring, but the water suddenly exploded and from the sea emerged the victorious hunter with his slain prey. The water was now empty. The bear waited for another opportunity, but the white whales had fled into the depths. The hunter was not prepared for bear, even though he would relish the contest if the bear had known of his presence. The intent now was not to hunt, but to return home and make final preparation for the coming season. In silence, he returned home to inform the hunters that the quest for whale would begin and that they must prepare.

The hunter rose early as dawn comes quickly with the coming of spring. He fed on the broth and meat of the white whale that was taken from the bear by other hunters. The whale was his as he had witnessed the kill, but the bear belonged to the hunter who had thrust the mortal wound.

The hunter tenderly put the ivory whale effigy that had overlooked his berth into his bag and left his home. In the tunnel awaited the rawhide lines and sealskin floats that were made weeks before. The clothing and the cover for their craft must be new; only the weapon and the effigy had seen other hunts.

The men worked quickly. The boat glided effortlessly over the ice and the snow. Soon they were at the water's edge where final preparation began.

A whale surfaced near the launch, and floated for long moments. The hunter lashed the effigy of the whale onto the bow of the craft, as other hunters attached the lines to the floats. The whale dove slowly and exposed its flukes to the hunters. In silence, the hunters knew the whale was telling them, "I am the first, and others follow."

No words could express the bond between man and whale, but they all knew that it was created by others long before them. The seasons had determined their meeting for many generations, a timeless process expressed by the hunter's unspoken thought, "We are descendant of the ancient hunter who pursued your ancestors, and the path of my people and your people will cross again. It has always been so." ■

Helicopter pilot Ferdie "Fritz" Wohlwend stands inside a stone-lined house depression at Skeleton Butte within the Bering Land Bridge National Preserve. (Jeanne Schaaf, National Park Service)

during Arctic Small Tool tradition times.

The Thule tradition, a maritime or sea-mammal-based economy, developed in the central and western Bering Strait area by 2,000 years ago, and includes the Okvik, Old Bering Sea, Birnirk and Punuk cultures. This tradition differed markedly from the Arctic Small Tool tradition and may represent the movement of new populations into the area or the adoption, perhaps in response to climate change, of a new technology by resident peoples.

Maritime hunters along Alaska shores became proficient at open-water gray and bowhead whale hunting by 1,000 years ago, but continued a mixed seasonal round including the taking of sea mammals, caribou hunting and fishing. Villages, camps and burial grounds of the Western Thule culture are known along the entire coast of the peninsula and in several interior locations. Today their descendents, the Inupiaq-speaking people of Wales, Shishmaref and several small villages continue to practice many of the same seasonal subsistence traditions. ■

Precontact Aleut Culture

By Douglas W. Veltre

Editor's note: *Dr. Veltre is a professor of anthropology at the University of Alaska Anchorage and has worked in the Aleutians and Pribilofs for many years.*

For at least 4,000, and as many as 9,000, years prior to Russian discovery of Alaska in 1741, Aleuts occupied the western end of the Alaska Peninsula, the Shumagin Islands and the Aleutian archipelago. Though related, Aleuts were linguistically and culturally distinct from their Eskimo neighbors to the east on the upper Alaska Peninsula, on Kodiak Island and on mainland southwestern Alaska. Like all other Alaska Natives prior to contact, Aleuts were a foraging people, depending on foods and resources obtained entirely by hunting, gathering and fishing. Also like other Alaska Natives, they were the descendants of peoples who migrated from Asia to North America some 10,000 to 15,000 years ago.

In describing Aleut lifeways as they existed prior to the arrival of the Russians, it should be realized that Aleuts have the unfortunate distinction of having had the longest and harshest history of contact with outsiders of any Alaska Native group. Such substantial changes occurred in their traditional precontact culture so quickly following initial contact, before their culture could be recorded, that it is nearly impossible to reconstruct with certainty many details of earlier Aleut lifeways. Knowing this, it is worthwhile to note the sources of information that contribute to attempts at ethnohistoric reconstruction. They include: archaeological data from excavations throughout the Aleutians; early Russian period travelers' and priests' accounts, maps and drawings; clothing, hats and other objects now in museums that were obtained in the early contact period from the Aleuts who made and used those objects; Aleut oral traditions; and aspects of modern Aleut lifeways that show continuities from the past. Each source has its own limitations in what it can offer, but the ethnohistoric research strategy pools all information for painting the best possible picture.

Establishing precisely how many Aleuts there were at the time of Russian contact is a nearly impossible task. Because the Aleut population declined so drastically and so rapidly following contact, and because it was only much later in the Russian period that accurate censuses were made, we have only archaeological data and early travelers' reports from which to derive our best estimates of the aboriginal population size. Although some figures go higher and some lower, it is reasonable to estimate that some 12,000 to 15,000 Aleuts lived in the region in the precontact period. Because the islands of the eastern Aleutians are generally larger and have more coastline available for settlement, the Aleut population was concentrated there, with fewer people living in the central and western islands.

Aleut life focused almost exclusively on the sea as the source of most food and fabricational

The name "Aleut" is of uncertain origin, though it likely was used by those people living in the Near Islands, at the western end of the Aleutian archipelago, to refer to themselves. Although it was not traditionally used by Aleuts to refer to themselves as a whole, early Russian fur hunters adopted the term in that manner; in fact, they indiscriminately applied the name to neighboring Alutiiq-speaking Eskimos as well. Today, change appears to be occurring, as some Aleuts have begun to employ the terms "Unangan" or "Unangas" — meaning "we, the people" in the eastern and central dialects of their language, respectively — to refer to themselves.

resources, and their subsistence economy was predicated on cooperation in obtaining many foods and on sharing those products within a large family sphere. Food items included marine mammals such as sea lions, harbor seals, sea otters, fur seals and whales; marine invertebrates such as sea urchins, clams, mussels, chitons and octopus; eggs and birds such as murres, puffins, ducks and geese; and fish, including ocean species such as cod and halibut and anadromous species such as Dolly Varden and several species of salmon. Plant foods, such as crowberries, wild rice and wild celery, provided only a small percentage of the Aleut diet. Aleuts hunted sea mammals and birds on the open ocean from superbly crafted kayak-style bidarkas; men hurled harpoons and spears at their prey with the aid of a throwing board for greater thrust.

In addition to their worth as food resources, most of the animals Aleuts hunted were equally valuable for the fabricational materials they provided. Sea lions are a good example. Their bones were used for various tools, including harpoon heads and digging tools; skins were sewn into boat covers; whiskers adorned bentwood visors; cleaned and split intestines provided material for rain gear; esophagus, stomach and intestines were made into a range of storage containers; and teeth were grooved so they could be suspended as pendants. Other animals were used in similar fashion.

Non-edible resources used solely as fabricational materials included driftwood, which supplied the main building material for Aleut houses; stone, such as basalt and obsidian, which was chipped into a wide range of projectile points, knives, adzes and scrapers; grass, especially beach rye, which was collected to be made into finely woven baskets, mats and other items; and ochre, an iron-rich, soft stone that was ground into a fine powder that served as an important paint pigment.

Among the more bountiful sites yet excavated in the Aleutians is the Chaluka midden at the modern village of Nikolski on Umnak Island. In this 1962 photo, the midden is buried under the cluster of buildings to the right of the village. The Chaluka midden represents about 4,000 years of Aleut occupation of this site. (Allen McCartney)

One distinctive aspect of the Aleut subsistence economy was that the diverse resources included some plentiful forms that were obtainable by Aleut males and females of almost any age. For example, most Aleuts could walk along the shore at low tide and collect sea urchins, chitons and other marine invertebrates. Likewise, Aleuts too infirm to

Waldemar Jochelson took this image of men from Nikolski village wearing gut kamleikas and bentwood visors and holding bone-headed spears and throwing boards during the 1909-1910 Aleut-Kamchatka Expedition sponsored by the Imperial Russian Geographic Society. Jochelson led the anthropological section of the expedition and excavated on Hog and Amaknak islands in Unalaska Bay. (Courtesy of Douglas W. Veltre)

venture out in boats could still fish for halibut, cod, salmon and other fish directly from shore. Thus, while on the one hand only younger men hunted sea mammals pelagically, on the other hand all members of an Aleut community could participate to a significant degree in feeding their families.

It is not surprising that because of Aleuts' reliance on the sea, their settlements were situated at the most advantageous coastal locations. Sheltered bays, stream mouths and spits were favored locations from which to hunt and gather food. As did Natives elsewhere in Alaska, Aleuts maintained both main villages, occupied by a larger number of people for much of the year, and seasonal subsistence camps, used by smaller groups for hunting and fishing. However, the abundance and concentration of food resources in the Aleutian region made it possible to maintain fixed village locations for long periods of time, something not possible in all areas of Alaska. For example, at the archaeological site of Chaluka, in the contemporary village of Nikolski, a record of essentially continuous Aleut occupation extends from 4,000 years ago to the present. Some residents of Nikolski still live atop this ancestral site.

Aleut villages were flexible in membership and variable in size, some consisting of only a family or two, while others may have numbered more than 200. Extended families lived together in semisubterranean barabaras, or *ulax,* houses built from driftwood, whale bone and stone and having roof entries. While most archaeologically known Aleut houses are small and have only a single room, some are quite large and complex. These houses, known only from the eastern area of Aleut occupation, had main living areas as large as 20 feet by 130 feet, with several semisubterranean side rooms attached to this through narrow passageways. Such structures could have been home to 100 people.

One area of uncertainty regarding precontact Aleuts is the realm of social organization, particularly the method by which Aleuts reckoned kin relations. The precontact kinship system changed completely before many details of it were recorded, and anthropologists are left to discern the underlying system from the few kinship terms recorded by early travelers. One possible interpretation holds that Aleuts possessed a matrilineal kinship system, similar in basic form to that well-known within traditional Tlingit culture. In such a society, the most important kin ties are traced by men and women through their mothers, not their fathers. Patterns of working, sharing and marriage were all influenced by one's membership in a particular matrilineal group.

Aboriginal Aleut society was also ranked, at least in the eastern Aleutians, with the highest ranking going to those individuals having the greatest wealth, the most slaves (Aleut and Eskimo war captives), the largest families, the most local kin support and the closest proximity to important subsistence resources. Signs of rank included finer dress and elaborate personal ornamentation. Tattooing and wearing of lip plugs, or labrets, were some of the means by which an individual could proclaim rank. Regional political alliances beyond the village level existed, though the manner in which they functioned is unclear.

The realm of Aleut ideology is not well-known. As elsewhere in Alaska, Aleut religion was clearly animistic, there being a general belief in the existence of spiritual aspects of humans, animals and inanimate objects such as rivers and mountains. Belief in the power of human spirits was nowhere more clearly evident than in the various means by which deceased persons were treated. While slain enemies were sometimes dismembered to release their otherwise powerful and dangerous spirits, deceased relatives were treated with great care and respect. Some were buried in laboriously constructed burials on hills close to a village, while others, likely those of highest rank, were occasionally mummified and placed with their belongings in caves along the shoreline.

Successful hunting was predicated on living harmoniously with the spirit world, including observing proper ritual behavior. Shamans served as magical-religious specialists, mediating between the everyday and spirit worlds. They were called on to cure the sick, ensure hunting success and bring victory in battle. Although much was never recorded before being lost, Aleuts had a rich oral tradition that included stories of ancestral heroes, myths of animal protectors and other supernatural beings, songs, narratives of everyday life and proverbs.

While other Alaska Natives adapted to life in the interior or along other coasts, Aleuts thrived in their island chain on the edge of the North Pacific and the Bering Sea. Employing often ingenious technology and a keen knowledge of the environment, developing social institutions that provided for mutual support through sharing and kinship, and integrating all aspects of life with a world view that placed people as but one part of a single natural and supernatural system, Aleuts lived successfully for thousands of years. Although much has changed in the last 250 years, many aspects of this heritage survive today among Alaska's Aleut people. ■

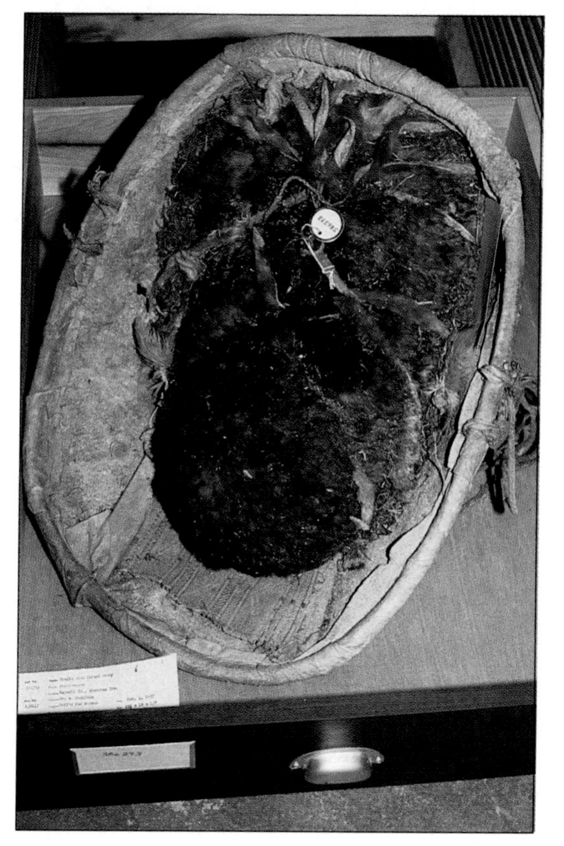

RIGHT: *The Smithsonian Institution's collection on Alaska prehistory includes this Aleut infant mummy bundle made of sea otter skins inside a skin cradle. (Douglas W. Veltre)*

FAR RIGHT: *Archaeological data derived from excavations is one of the ways scientists have reconstructed Aleut prehistory. Here Allen McCartney, professor of anthropology at the University of Arkansas in Fayetteville, excavates part of a settlement that was occupied until the early 1800s on Unalaska Island. The islands of the eastern Aleutians are larger, prompting early people to occupy these islands more densely than they did islands farther west. (Douglas W. Veltre)*

Prehistoric People of Alaska's Interior

By David R. Yesner and Kristine J. Crossen

Editor's note: *Dr. Yesner is on the staff of the University of Alaska Anchorage's department of anthropology; Professor Crossen is on the staff of the geology department.*

The bluff tops along the river valleys of interior Alaska, north of the Alaska Range, particularly the valleys of the Tanana River and its tributary, the Nenana, are slowly yielding evidence of 12,000 years of occupation by prehistoric Alaskans. The glaciers of the last ice age eroded and ground the rocks of the Alaska Range, producing a glacial silt called loess, which was transported by wind and water down these river valleys, and finally blown on top of the bluffs that surround the valleys. These loess deposits contain evidence for the first Alaskans in the form of artifacts, fireplaces and other debris associated with buried soil horizons or paleosols. In the 1960s, John Cook and Robert McKennan found evidence for projectile points, blades and other tools more than 11,000 years old at Healy Lake in the southern Tanana Valley. In the 1970s and 1980s, attention turned west to the Nenana Valley, where evidence for similarly old occupations was found at a number of bluff-top sites excavated by Roger Powers and his students at the University of Alaska Fairbanks. The Nenana Valley sites are now well-known, and include Dry Creek, Panguingue Creek, Walker Road, Owl Ridge and Moose Creek, all of which are older than 11,000 years.

Beginning in 1990, attention once again shifted to the Tanana Valley with the discovery of the Broken Mammoth site north of Delta Junction. The Broken Mammoth site was unique in that not only was evidence of early stone tools preserved, but also bones and other organic materials that are not normally preserved in the acidic forest soils of interior Alaska. Bone at Broken Mammoth is preserved because the loess is derived from limestone rocks, and because it is more than 6 feet deep, which has prevented modern soil acids from leaching down into the late ice-age deposits and destroying the organic materials.

The name of the site, Broken Mammoth, comes from the discovery by Charles Holmes of the Alaska Office of History and Archaeology of mammoth tusk fragments eroding from the base of the loess cap, 80 feet above the Tanana River. With the mammoth ivory were found fragments of stone tools, charcoal from camp fires and bones of large and small mammals as well as birds. These finds encouraged the development of a field school by the University of Alaska Anchorage at the site, as part of a joint project with the Office of History and Archaeology. During the next four years, much more evidence of tools, bones and other materials came to light.

The excavations at the Broken Mammoth site, as well as at the nearby Mead and Swan Point sites, indicated that these sites were re-occupied at least four times: between 11,000 and 11,800 years ago; around 10,000 years ago; around 7,500 years ago; and between 3,000 and 4,000 years ago. Many of these

Economically Important Fauna from the Broken Mammoth Site

Source: Dr. David R. Yesner

Bison
Elk (wapiti)
Mammoth
Caribou
Dall sheep
Camelid (?)
Arctic fox
Lynx
River otter
Muskrat
Porcupine
Beaver
Marmot
Hare
Ground squirrel
Red squirrel
Tundra swan
Geese
Ducks
Ptarmigan
Fish

Students and archaeologists excavate the Broken Mammoth site at the top of this bluff in the Tanana River valley near Delta Junction. A particularly thick cap of loess helped preserve the artifacts at Broken Mammoth. (David Yesner)

dates were derived by the new accelerator mass spectrometry (AMS) radiocarbon dating method. The first two occupations represent the record of some of the earliest Alaskans, contemporaneous with the earliest people of the Nenana Valley and possibly the Mesa people of northern Alaska. Sediments from the site tell us that a stable land surface existed during the time of the late ice-age occupants, but both before and after that time the region was dominated by a windy and arid climate that was probably inimical to human occupation. The sedimentary record also indicates that the Tanana River probably meandered back and forth across the valley, at times closer and at times farther from the site.

How did the first Alaskans live? Unfortunately, the bluff-top campsites that archaeologists have uncovered in the Tanana and Nenana valleys probably only represent a portion of the overall yearly settlement pattern. They represent places where some portion of the total hunting band, possibly groups of extended families, camped for some weeks or at most a few months, observing movements of game animals in the valleys below, retrieving and butchering the animals that they killed, and then transporting the butchered carcasses back to their main village. In the meantime, they made stone, bone and ivory tools, sewed and repaired skin clothing, built camp fires and possibly simple teepeelike hide-covered dwellings, and lived off a portion of their kill. The main villages may have been down in the river valleys, where they have been wiped out by 10,000 years' movement of the Tanana and Nenana rivers.

Support for these statements comes from the discoveries of artifacts and evidence of other activities at the Broken Mammoth site. The artifacts include stone projectile points similar to those known from other northern Paleoindian sites; a mammoth ivory projectile and associated spearthrower handle,

LEFT: *The University of Alaska Anchorage, in cooperation with the Alaska Office of History and Archaeology, has operated a field school at Broken Mammoth, where excavation has yielded remains of hearths, fragments of tools and pieces of bone from ancient animals. According to radiocarbon dates, the site has been occupied at least four different times beginning between 11,000 and 11,800 years ago. (David Yesner)*

ABOVE: *This mammoth ivory projectile point, painted with red ocher, was unearthed at Broken Mammoth. Red-ocher-painted artifacts have been found at several sites in Alaska and scientists associate this decoration with the spiritual life of ancient people. (David Yesner)*

similar to some known from Siberia and the contiguous United States; stone tools for butchering game and processing hides and bones, such as knives, scrapers and burins; and an eyed, bone needle and a bone toggle for tailored skin clothing. Activity areas present at the site include large workshop areas for the manufacture of stone tools from local quartz ventifacts and Tangle Lakes chert; well-formed stone hearths; and possible dwellings, indicated by circular discard patterns of tool and bone debris.

The bones preserved in the Broken Mammoth site have done much to help scientists reconstruct the environment and lifeways of the first Alaskans, and to tell how these changed through time in interior Alaska. In later time periods, for example, more aquatic animals such as beaver, river otter and muskrat appeared, as well as animals of more forested environments, such as red fox, red squirrel and porcupine. This reflects the somewhat wetter and more forested environment present in the area today. But for the late Pleistocene period, between about 12,000 and 10,000 years ago, the lifestyle of the earliest inhabitants centered around four types of animals: large game, principally bison, elk and caribou; small game, principally snowshoe hare, ground squirrel and arctic fox; birds, principally ducks, geese and swans; and a small salmonid fish, possibly grayling. Considering that until recently archaeologists had depicted the first Americans exclusively as big-game hunters, the diversity of subsistence reflected at the Broken Mammoth site is remarkable. Both the bison and elk are of the large ice-age variety, larger than modern species. The evidence of bird hunting and fishing is the earliest for anywhere in Alaska, and tells us that these important resources were well-established by the end of the Pleistocene. The earliest occupation of the Broken Mammoth site contains more bird remains, while the later, 10,000-year-old occupation contains more mammal and fish remains. These patterns probably reflect the use of the site at different seasons. The bison and elk were most likely taken in the fall and winter, when herds would have moved to lower elevations in search of forage along wind-cleared river valleys. Waterfowl were most likely hunted during spring and fall; the tundra swan, in particular, inhabits the area during the late summer and fall. Fish would probably have been taken in summer, while other species would have been available year-round.

The mammoth remains from the Broken Mammoth site consisted only of mammoth tusks. Why other mammoth bones were not present is somewhat of a mystery. Two explanations are possible: first, early hunters may have butchered the mammoths in river valleys where they boned them out, like some elephant hunters do today. Following this reasoning, if they took parts of the mammoths back to their camps, it was only the meat, not the bones. Another possibility, perhaps more likely, is that the mammoths had become extinct in Alaska shortly before the first hunters came. If so, these mammoths may not have been encountered again until these hunters veered southward toward the plains of Alberta, Montana, Wyoming and Colorado.

The Broken Mammoth site tells us mostly about hunting, fishing, toolmaking and hideworking, the subsistence-related and domestic chores of early Alaskans. But there are also brief glimpses of spiritual life. For example, the mammoth ivory projectile was painted with red ochre, a substance that had spiritual meaning, which may suggest that important ritual surrounded the manufacture and use of these implements. Perhaps more such evidence will come to light in future excavations at the Broken Mammoth site. ■

BELOW: *This eyed bone needle was found at the Broken Mammoth site. (David Yesner)*

RIGHT: *Working carefully, an archaeologist scrapes away loess from leg bones of an elk. (Richard Vanderhoek)*

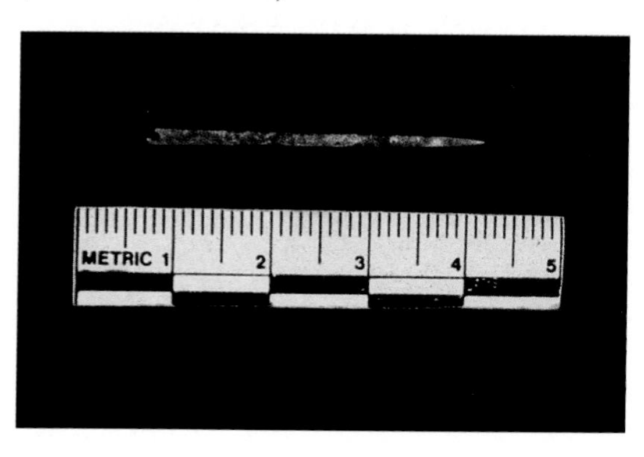

A Prehistory of Southeast Alaska

By Wallace M. Olson

Editor's note: *Dr. Olson is professor of anthropology emeritus, University of Alaska Southeast. He is the author of* The Tlingit: An introduction to their culture and history *(1993) and* The Alaska Travel Journal of Archibald Menzies, 1793-1794 *(1993).*

■ *History of Archaeological Research*

The first systematic studies of southeastern Alaska's prehistory were made by Dr. Frederica de Laguna and her associates at Angoon, on Admiralty Island, in the summers of 1949 and 1950 and at Yakutat in 1952 and 1953. The results were published as *The Story of a Tlingit Community: A problem in the relationship between archaeological, ethnological and historical methods* (1960) and *The Archaeology of the Yakutat Bay Area, Alaska* (1963). These two reports set the stage for future research and are still considered standard references for the area.

In 1965, while excavating a beach ridge at the Ground Hog Bay 2 site near Excursion Inlet, Robert Ackerman, of Washington State University, and his team made an important discovery. He writes that when they found a small stone microcore from which microblades had been struck, "We knew then that we had broached the prehistoric stage of the northern Northwest Coast cultural continuum." Lower cultural layers at Ground Hog Bay have been dated from 10,180 ± 800 years ago and extend up to early historic times. Then in 1985, at Locality 1 at the Warm Chuck Lake site on Heceta Island, Ackerman excavated a shell midden dated at about 8,200 years ago, which contained stone tools and provided the first reliable look at the diet of these early hunters and gatherers.

In March 1978, while excavating a site at Hidden Falls on Baranof Island, Forest Service archaeologist Stan Davis also found stone tools that were dated to about 9,500 years ago. Like Ground Hog Bay, the Hidden Falls site has several later components ranging in age from 5,000 years old up to the present.

Following these early studies, Madonna Moss and John Erlandson of the University of Oregon, Chris Rabich Campbell of Ketchikan, Herb Maschner of the University of Wisconsin at Madison, Charles Holmes of the state Office of History and Archaeology and Hiroaki and Atsuko Okada of Hokkaido and Tokai universities in Japan, have also made significant contributions to our knowledge of the prehistory of Southeast. In addition, U.S. Forest Service archaeologists John Autrey, Karen Iwamoto and Mark McCallum continue to provide new data from their work in Tongass National Forest.

■ *Linguistic Clues*

In some cases, linguistic analysis and comparisons can furnish prehistorians with information about relationships between various societies. Along the North Pacific coast there are families of languages with completely different words, sounds and sentence structure. On the northern end, the Eskimo languages and Aleut are distantly related, but these languages are altogether distinct from Athabaskan and Eyak, which are related. On the southern end are the Salishan, Wakashan and Penutian families of languages. In between we find the Tsimshian, Tlingit and Haida languages that have with no linguistic relationship to each other except some borrowed words.

Although Tlingit is a linguistic "isolate" -- unrelated to any other language -- its grammar, especially verb construction, is similar to Athabaskan. But the sounds of Tlingit and Athabaskan vary, and outside of a few loan words, the vocabularies are completely different. Linguists are still unable to link Tlingit to any other specific language.

People often ask archaeologists if a prehistoric site is "Tlingit" or "Eskimo" without realizing that what they are asking the researcher to do is to look at the tools and remains and tell them what language the people were speaking when they occupied the site. In most cases, there is no way the archaeologist can link tools, shell middens and subsistence patterns to a particular language. With sufficient human skeletal material, archaeologists can sometimes relate the artifacts to a biological population and a language group, but in Southeast there are only a few archaeological sites with any human remains, and these remains are so recent or fragmentary that they cannot be used to identify early inhabitants of the area.

In a 1991 paper, Stan Davis proposed an hypothesis using archaeological data, information on the Athabaskan, Eyak and Tlingit languages and research on dental patterns. He suggests that in the past an extensive cultural network linked the inhabitants of Kodiak Island, the Alaska Peninsula, Cook Inlet, Southeast and the Pacific Northwest coast. Jeff Leer, an expert on the Tlingit language, has looked at Tlingit in relationship to other coastal languages, and he too suggests that at some time in the past the speakers of Aleut, Eyak and Haida were in regular contact and that Tlingit may be a later intrusion into the network.

All of this means that on the North Pacific coast, while archaeological and ethnological studies show cultural sharing and similarities, there is a great deal of linguistic diversity. Madonna Moss recommends that we look beyond the artificial boundaries of language and "culture areas" such as Eskimo, Aleut and Northwest Coast and focus more on the similarities these people share because of their maritime way of life and mobility.

Several prehistoric sites cluster near Warm Chuck Lake on Heceta Island. Locality 1, on the small hill just this side of the lake at upper left, is a small shell midden about 8,000 years old. Locality 3, a deep shell midden about 5,000 to 6,000 years old, is at the bottom of a cluster of trees on the lower right shore of the lake. The village on Warm Chuck Inlet, at lower center, was occupied up to about 1900 and may be more than 1,500 years old. (Wallace Olson)

■ Origins

For years, prehistorians have speculated on the origins of the first inhabitants of Southeast and the northwest coast of North America. There is little doubt that the first Americans emigrated out of Asia, but the time and routes

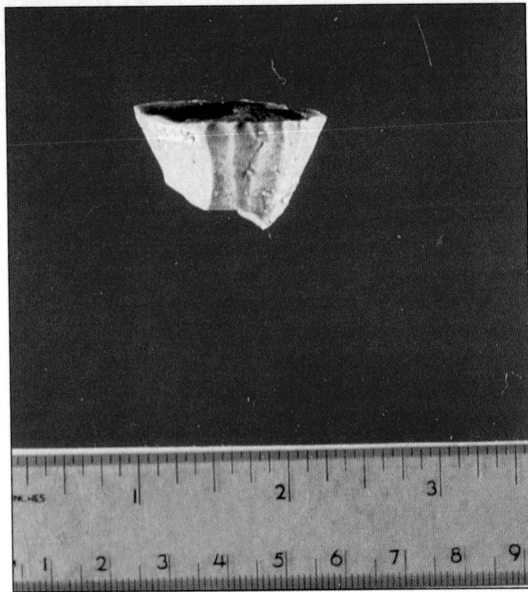

LEFT: *Smoking has enhanced the photographing of the frontal, wedge-shaped flaking of this obsidian microblade core found at the Ground Hog Bay 2 site near Excursion Inlet. Known sources of obsidian for prehistoric people who lived about 10,000 years ago in Southeast were Mount Edziza in interior British Columbia and Sumez Island west of Prince of Wales Island in southern Southeast. (Robert Ackerman)*

LOWER LEFT: *This tiny chert microblade core recovered in 1965 revealed a new dimension in the prehistory of the southeastern Alaska. An archaeological team was excavating a beach ridge when they uncovered the artifact. Radiocarbon dating of material found at the ridge indicates that some items are about 10,000 years old. (Robert Ackerman)*

of migration are still in question. By 10,000 years ago the last ice age was ending, and people would have been able to travel along the coast in small watercraft. In recent years, an increasing number of prehistorians are looking at the possibility of coastal migrations by some of the earliest Alaskans.

Because the Hidden Falls and Warm Chuck Lake sites are on islands, and the food remains at Warm Chuck Lake Locality 1 indicate an emphasis on bottom-dwelling fish, it's clear that these early settlers must have used some type of watercraft for transportation. The obsidian used at Ground Hog Bay and Hidden Falls appears to have come from Mount Edziza, inland in British Columbia, and Sumez Island, west of Prince of Wales Island, indicating that by 10,000 years ago the people of Southeast were already engaged in long-range trade and travel.

Ackerman points out similarities between tools from the lower level of Ground Hog Bay 2 and artifacts from the Denali complex of interior Alaska and possibly with Siberian technology. These tool kits contain microblades and microblade cores, bifacial tools, burins, scrapers, notched stones and choppers. The microblade cores are wedge-shaped with blades removed from the frontal portion. The lowest component of the Hidden Falls site and Locality 1 at the Warm Chuck Lake site contained comparable tools. Don Dumond, of the University of Oregon, lumps the Alaska sites with frontal, wedge-shaped microblade cores whose striking platforms were formed by the detachment of platform tablets and refers to these sites as the "Paleo-arctic tradition." The shell midden at Warm Chuck Lake Locality 1 contains a large percentage of butter and littleneck clams, mussels, fish bones and a few sea mammal bones, showing that the people there were already adapted to a marine way of life. A large percentage of the fish bones are from species other than salmon, indicating that they were probably doing a lot of offshore fishing. This earliest coastal cultural tradition seems to have persisted from 10,000 years ago to about 5,000 years ago. However, since the microblade cores at Warm Chuck Lake are no longer frontal and wedge-shaped, but are blocky with blades removed from the lateral faces and from the entire circumference of the core, and are distinct from cores from the interior but similar to a type widely distributed on the Northwest coast, Ackerman maintains that by 8,200 years ago, the "northwest coast had its own cultural dynamic."

So far, no skeletal remains of these early inhabitants have been found, but recently an extensive system of caves has been discovered on Prince of Wales Island and if the caves contain any human remains, we may be able to determine the biological relationships of these people.

These earliest inhabitants were probably living in small groups, traveling in boats, canoes or kayaks, living off the rich coastal resources and leaving behind little more than a few traces of their passage.

■ *Transitional Stage*

Beginning about 6,500 to 5,000 years ago, a change in tool technology appears in the archaeological record, with a shift from flaked to ground stone and bone tools accompanied by technological diversity, increased settlement size and perhaps the start of more complex social systems. Davis calls this the Transitional Stage between the earlier coastal cultural tradition and the development of the Northwest Coast tradition. Whether the culture evolved locally or diffused from some other source has not yet been determined. But it may be that it was at this time that a network developed between the Aleuts, Eyak and Haida.

Few sites from this period have been studied, but in 1990 at Warm Chuck Lake Locality 3, about 1,000 feet from Locality 1, a team led by Hiroaki and Atsuko Okada excavated a 4-foot-9-inch-wide trench through what appears to be an extensive shell midden, radiocarbon dated to 5,520 ± 300 years ago. At the point where the trench was cut, the midden was 26 feet wide and had an average depth of 12 to 16 inches. At the deepest point, the shell deposit was 7.6 feet deep. Using soil probes, John Autrey estimates the midden to be between 165 and 330 feet long. The only artifacts recovered were 16 pieces of worked bone and eight stone artifacts including a large, bilaterally barbed bone harpoon and what appears to be a fragment of a big fish hook or some kind of ornament. In addition to a massive amount of sea shells, the excavation produced bones from birds, sea mammals, one whale vertebrae and a few remains of land animals. Although they probably used local berries, roots, seaweeds and bird eggs for food, remains of these resources do not preserve well and were not found in the midden. The heavily forested ridge above the midden, which may contain a campsite or settlement, was not excavated.

This transitional site shows that the earlier coastal subsistence pattern persisted through several centuries and was able to support a substantial population. Even today, Natives gather a great deal of food from the shoreline. With a dependable food supply, it is likely that the human population continued to increase as well. And since the midden is quite large, their settlements may have been larger also.

■ *The Development of Northwest Coast Culture*

Based on tool technology, faunal remains and evident changes in settlement patterns, Davis divides the next period into the early, middle and late development of the northern Northwest Coast culture.

The artifact inventory at the early developmental level, from 5,000 to 3,000 years ago, includes ground stone points, small adzes and abraders, unilaterally barbed harpoon heads and labrets. But microblades, bifacial tools, flaked stone points and burins or engravers are no longer present in the sites, telling us that there was a change in technology or population or both. The presence of labrets is provocative because labrets were also used by Aleuts and Eskimos and Natives farther south on the Northwest coast, but not by interior Indians. The labrets are perhaps another clue to some earlier cultural sharing between people of diverse language families. If Athabaskans did not use labrets, and the Tlingit language shows some grammatical similarities or relationship to Athabaskan, how can we explain this linguistic affinity but cultural difference?

Farther south, along the British Columbia coast, there are indications of population increase and greater cultural complexity. Ackerman says that "The widespread intensity of coastal settlement of the British Columbia coast by 5,000 B.P. seems to reflect the dawning of a new social and economic order based on the elaboration of technology, trade, social structure, art, and ceremonialism." Looking at Hidden Falls material in Component III, dated at 3,000 to 1,300 years ago, Stan Davis points out that at this time, in addition to ground stone tools and mauls, there are composite toggling harpoons, unilaterally barbed bone points, labrets, incised

Japanese archaeologists work in the trench cut through the shell midden at Locality 3 at the Warm Chuck site on Heceta Island. (Wallace Olson)

Prehistoric people living on the west coast of Prince of Wales Island built elaborate fish traps and weirs to catch fish. These archaeologists stand amid a collection of posts that at one time anchored some of these fish traps. The traps and weirs were usually built at the mouths of streams. At high tide the fish could move upstream, but when the tide went out, the fish were stranded behind barricades or caught in basket traps when they headed downstream. (Chris Wooley)

bones and stones and drilled mammal teeth. Besides the diversity in tools and ornaments, there are signs of more permanent structures, and in addition to the fish and shellfish material, remains of both sea and land mammals are found.

During this time, increased reliance on salmon is indicated by the use of fish weirs. The Favorite Bay fish weir at Angoon, described by Madonna Moss and Jon Erlandson, has been dated between 3230 ± 80 and 2170 ± 50 years ago. This weir shows that mass harvesting of salmon was important in this developmental stage and suggests a gradual trend toward a greater reliance on salmon continuing into historic times. Steve Langdon and Chris Wooley have also identified many stone fish weirs in the region. On the other hand, Herb Maschner, in his doctoral dissertation, says that in Tebenkof Bay there was a "rapid change from a generalized open-water fishing strategy to one specialized on salmon after 700 bp."

Maschner proposes that somewhere in this developmental period, social and environmental conditions contributed to increased concern with rank and prestige in the social system whereby certain individuals could manipulate the situation to put themselves in a superior position. This stratification of society can be seen in the appearance of large houses and special burials from this period. For instance, in the settlement sites of this time, one house is often larger than the others, which Maschner interprets as the residence of the leading family or kin group in the community. Farther south on the coast, sites of equal age contain some of the early examples of animal crest designs that later become so prominent in Northwest Coast art.

What Davis calls the Late Phase of the Developmental Stage runs from about 600 to 200 years ago and is marked by the addition of stone bowls and lamps, harpoons with lashing holes and metal tools. Beautiful stone adzes, hammer stones, and mortar and pestles made by pecking and grinding, were still in use when the first Europeans arrived. This may be the time when large plank clan houses, dugout canoes and bentwood boxes came into use, and many of the tools were probably designed for woodworking.

Although the sites contain the remains of many sea mammals such as harbor seals, porpoises and sea otters, land mammals, including mountain goat, bear, beaver, marmot and muskrat are also represented. Maschner sees this period as a major turning point in the region's prehistory with a dramatic shift in diet, saying, "There is a total replacement of cod and herring with salmon, while seal and otter are replaced by deer." Why this shift in diet? Perhaps there was an increase in warfare, raiding and slavery.

From excavations in Prince Rupert harbor in British Columbia, it appears that by 3,000 years ago there was organized warfare on the Northwest coast. According to Moss and Erlandson, by 1,500 years ago Tlingits were using fortified sites for protection, and based on radiocarbon dates from these structures they concluded that "between A.D. 900 and A.D. 1400 there was a dramatic increase in the number of forts in use." In historic times there are many reports of raiding and warfare along the North Pacific coast, especially among the Tlingit, Haida and Tsimshian. It should be noted that the dates listed by Moss and Erlandson appear to be the earliest, not the latest dates, that the forts were occupied. For instance, the radiocarbon dates for the Rancheria site at Craig are given as A.D. 1041 to 1241, but in 1779, when Commander Arteaga visited this fort, he reported that "we found a great number of houses, built entirely of good Wood and freshly cut, and some parts were painted red and black, which Wood must have been very difficult to set up, so heavy were the timbers and crossbeams."

Based on his study of 150 sites in Tebenkof Bay and comparable sites in the area, Herb Maschner points out that in the "Middle Phase" there are large shell middens with an emphasis on herring and cod fisheries with small villages located on "convoluted coastlines with excellent intertidal resources" that are "approximately the size of the area inhabited by a single lineage house in the Late Phase villages." But in contrast, the Late Phase "villages are located on long, straight shorelines with excellent long distance views of the open water around them." He goes on to say that all of these changes "occur exactly when it is hypothesized that the bow and arrow arrives on the Northwest Coast." His explanation is that with increased warfare using the bow and arrow, offshore fishing became more dangerous, and the smaller villages joined together into the larger winter settlements seen in historic times. He continues: "I argue that the increased efficiency of this weapon of war stimulated lineages to move into large social constructs [villages] for defensive reasons. These larger groupings allowed striving leaders of the largest lineages to take political control and maintain it for the benefit of themselves and their immediate kinsmen." The maps produced by Walter Goldschmidt and Theodore Haas in 1946 showing the historical Native use and occupation of Southeast seem to indicate that in the not too distant past, the people were living in smaller, scattered settlements and only recently had moved into larger communities like Hoonah, Kake and Angoon. But whether or not the people moved into larger, centralized villages for the reasons given by Maschner is still in question. For instance, Ackerman says that the movement of Hoonah to its present location was to get better anchorage for large vessels. The location of other modern villages may be due to missionary influence, presence of a cannery or because it was a good place for commercial fishing. In any case, there seems to be a regular pattern of cultural evolution in subsistence, technology, social systems and art in Southeast from 5,000 years ago into historic times.

In many respects the archaeology of this region is still in its infancy. As a temperate rain forest, with heavy vegetation, the Tongass National Forest is a difficult place in which to find ancient sites and carry out careful archaeological excavations. Although we now have hundreds of known sites and radiocarbon dates, the more than 45,000 square miles of Southeast probably contain

These argillite and andesite cores and choppers represent a collection of artifacts found by Dr. Robert Ackerman and his team near Excursion Inlet. (Robert Ackerman)

Chilkat blankets, like the one worn here by a dancer in Haines, have been part of the Tlingit culture for hundreds of years. (Staff)

thousands of other sites that will no doubt someday yield new information about the prehistory of this region.

■ *Protohistory and Oral History*

The term "protohistory" is used to refer to the transition between prehistory and the historical record. In Southeast, important changes were occurring during the protohistorical period, that is, at the time the first Europeans arrived. For instance, in the mid-18th century, Haida from the Queen Charlotte Islands began to migrate to Prince of Wales Island. To the north, Tlingit were expanding into Eyak territory and raiding into Prince William Sound and the Kenai Peninsula. The botanist Archibald Menzies noted that at Sitka and Angoon, the Tlingit were cultivating tobacco but the source of the tobacco is still unknown. When the Spanish first contacted the Tlingit and Haida in 1779, the Natives were familiar with iron and sought it in trade, but where did they develop this familiarity? There seems to have been too much iron on the coast for it to have come solely from isolated shipwrecks. Based on recent metallurgical studies of Tlingit and Haida objects at the British Museum of Mankind, Jonathan King says that the Natives had become skilled metal workers by the time these objects were collected. Where did they learn those skills? During this period the Raven's tail and Chilkat blankets were in use and the people were carving large house poles and canoes. How did these and other art forms evolve? It is clear that a great deal of research still needs to be done on the protohistoric period of this region.

The Tlingit and Haida have their own oral history, but in these accounts, events are almost impossible to date unless they can be associated with some natural phenomenon such as a volcanic eruption or glacial expansion or can be tied to an archaeological excavation. Ethnohistorians tend to question the historical accuracy of oral histories that claim to describe events that took place hundreds or thousands of years ago because stories can become garbled during many generations of retelling. However, totem poles, house poles, carvings, petroglyphs and songs may serve as physical reminders of past events, thereby allowing the Tlingit and Haida to preserve their history with greater accuracy.

George Emmons, John Swanton, de Laguna and Ronald Olson have recorded abbreviated accounts of Tlingit history and migrations, but in most cases clan histories are considered family property and are used for a variety of social, economic and political purposes. For the Tlingit, Haida and Tsimshian, nearly all history is clan history. Like all histories, Native or non-Native, written or unwritten, the selection of events reported and the interpretations of these events may vary, so it is impossible to say that something is "the Tlingit history of the Tlingit people." However, certain patterns emerge that indicate that the Tlingit migrated into Southeast from elsewhere, but where they came from varies from clan to clan. For instance, de Laguna writes that the Tlingit of Yakutat moved down from the north over the glaciers while Ronald Olson has them coming from the south. Some Tlingit say that their ancestors came to the coast by passing underneath a glacier.

In 1989, Chris Rabich Campbell wrote an exceptionally good article concerning the NexA'di clan that did not seem to fit into the Tlingit two-part or moiety system. She took earlier written accounts and combined them with interviews to produce an account showing that the NexA'di are an ancient clan with their own specific history as a third segment of Tlingit society. Even at this late date, she was able to reconstruct another small piece of Southeast prehistory.

Although prehistorians have learned a lot about the prehistory of Southeast during the past 50 years, it is clear that much work still needs to be done to combine Native oral history with the archaeological record, the very problem de Laguna began investigating nearly 50 years ago with her classic study of Angoon. ■

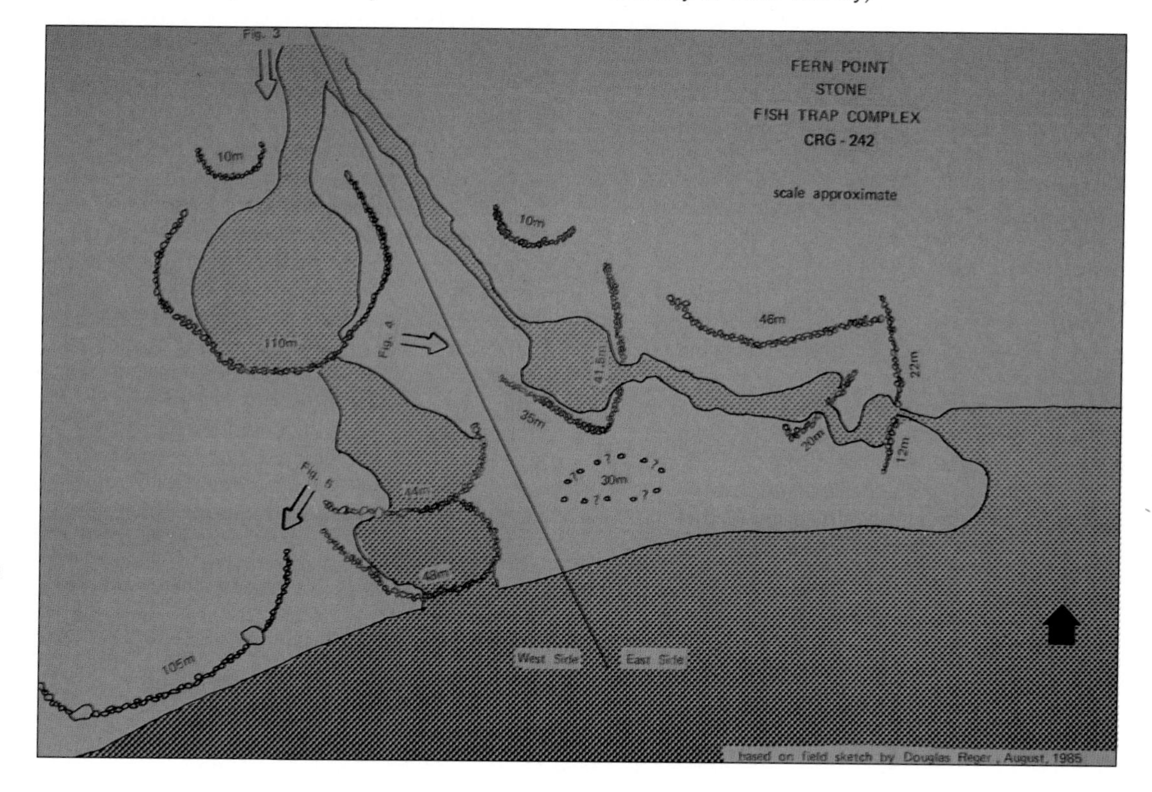

This diagram of a prehistoric site on Prince of Wales Island illustrates how stone walls were built to trap fish at different stages of the tide. (Alaska Office of History and Archaeology, courtesy of Chris Wooley)

Molecular Evolutionary Genetics of Indigenous Northern Populations

By Dr. Gerald F. Shields

Editor's note: *Dr. Shields is a professor with the University of Alaska Fairbanks' Institute of Arctic Biology. Some of the data on which this article is based were produced by his students Andrea Schmiechen and Kristen Hecker and his research collaborators Drs. Mikhail I. Voevoda of Novosibirsk, Russia, and Ryk Ward of the University of Utah, Salt Lake City. Dr. Shields acknowledges the technical assistance of Judy Gust.*

Most scientists think that humans first entered the New World from Asia and that they used the exposed Bering Land Bridge to arrive in North America. However, important questions remain regarding the pattern and process of human habitation of the Americas. For example, when did the first Americans arrive? Was their establishment in the Americas a continuous process or did various tribal groups move across Beringia in temporally isolated waves of closely related immigrants? If so, when did these migrations occur? Also, where in Asia did these early pioneers originate and is there evidence of their genetic legacy in the genes of Native peoples of the Beringian region?

Traditionally, archaeological descriptions of ancient remains have been used to date periods of occupation. The majority view among archaeologists is based on the thought that there is no firm archaeological evidence for a human presence in the New World prior to the so-called "Clovis arrow and spear point barrier" of 12,000 to 13,000 years ago. Nonetheless, there are numerous yet controversial claims for human habitation in the New World beginning as far back as 40,000 to 45,000 years ago. Linguistic affiliation of Native groups has been used to imply cultural diversity and hence period of habitation. However, the extent of diversity in Native American languages is unresolved. One authority describes three major groups of languages in the New World: Eskimo-Aleut, Na-Dene and Amerind. Others argue that it may have taken as long as 35,000 years for the linguistic diversity of the New World to develop.

No doubt these debates will continue as additional archaeological remains are discovered and as more refined linguistic analyses become available. There is, however, a different methodology that can provide an independent and objective perspective to these debates. This methodology employs modern molecular techniques to genetically type present-day Native groups from Beringia and adjoining areas and assumes that their genetic composition reflects their past genetic heritage. Such studies began in my laboratory at the Institute of Arctic Biology at the University of Alaska Fairbanks in 1988 and have involved undergraduate and graduate students as well as collaborations with geneticists in Russia and the University of Utah. Our group employs a new technique that allows recovery of sufficient quantities of DNA so that the linear sequence of a gene can be described from very small amounts of tissue. We study a segment of DNA transmitted through the mitochondria, small cellular inclusions whose primary function is to generate energy for cells. This segment of DNA incorporates mutation relatively rapidly and is thus a useful genetic marker among closely related Native groups.

Since mutations in mitochondrial DNA occur at relatively steady rates, this DNA can be used as an approximate molecular clock that accumulates genetic variation during the time since two individuals last shared a common ancestor. This means that large genetic differences between individuals correlate with long periods of separation from a common ancestor; little genetic divergence correlates with a shorter time period. We have reasoned that these studies of northern groups must include not only indigenous peoples of the Bering Strait region (Siberian Eskimos, Athabaskans, Aleuts, Yup'ik and Inupiat Eskimos) but also groups of eastern Siberia (Chukchi, Yukaghir and Altai), the Arctic (Greenlandic Eskimos) and the Pacific Northwest (Haida, Nuu Chah Nulth, Bella Coola and Yakima). Accordingly, our genetic studies include representatives of all the major linguistic groups of the region.

A major finding of our work indicates that populations of the northern regions of the New World (e.g. Athabaskans, Greenlandic Eskimos and Haida) possess only about half the genetic diversity of either the Amerindian groups (Nuu Chah Nulth, Bella Coola and Yakima) or those presently living in Siberia (Siberian Eskimos, Chukchi, Yukaghir and Altai). Low genetic diversity within and among New World northern groups suggests that these populations are younger than Amerindian or Siberian groups. Accordingly, we predict that the Amerindian radiation has occurred during the last 12,000 to 13,000 years while that of the northern New World groups has taken about half as long. These observations based on molecular data do not conflict with dates estimated from archaeological data.

There are two categories of DNA types among Yukaghir, Chukchi and Siberian Eskimos: those that are exclusively Old World and those that are shared between the Old World and the New World. This implies that the evolutionary affinities of Natives of northern North America extend west to the indigenous populations of eastern Siberia. They do not, however, include the Altai of central Siberia, who are themselves diverse and clearly distinct from all other groups we have studied.

Since low genetic diversity of northern New World groups would impact the distribution of genetic susceptibility to such common diseases as diabetes, rheumatic disease and heart disease, such a result could have considerable implications for public health. Moreover, models for health care structured on Amerindian populations may not apply to northern groups.

A second finding of these genetic studies is that there is little correlation between the languages the Native groups speak and the types of DNA they possess. For example, populations that speak Na-Dene languages

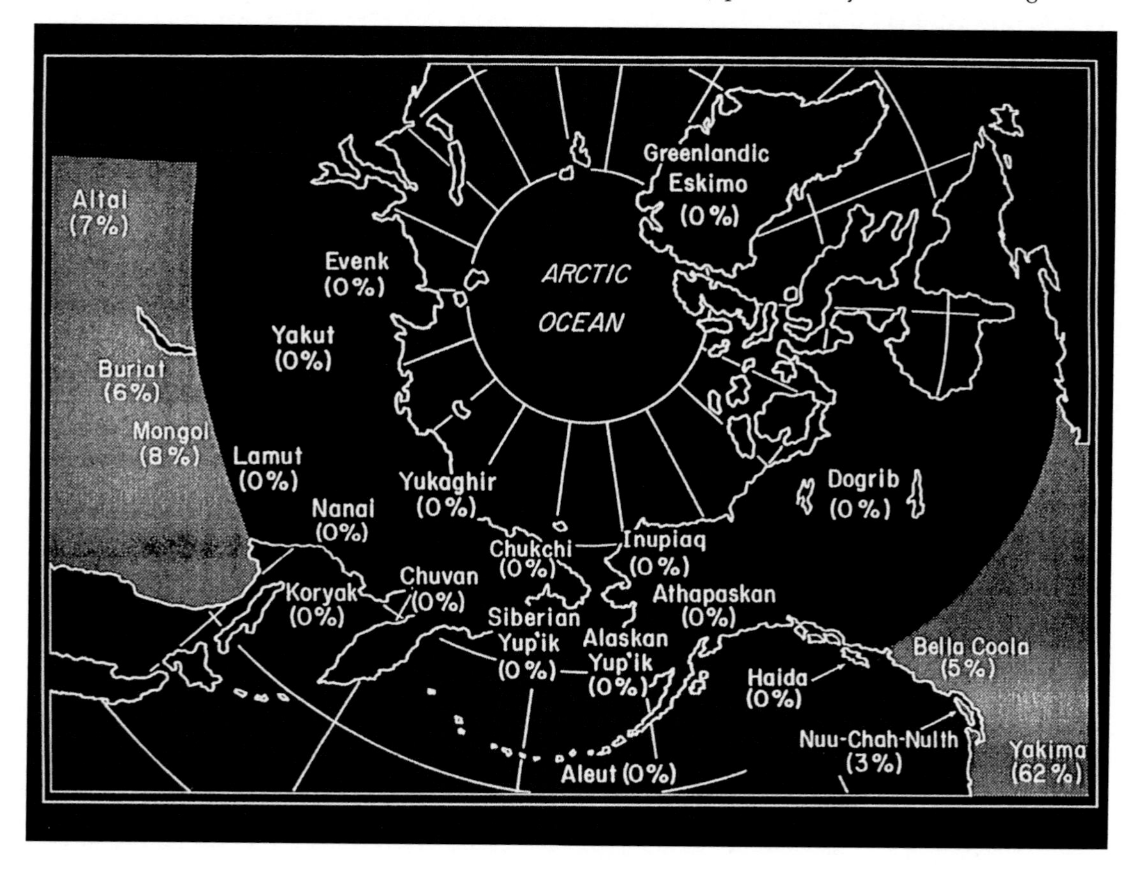

This map shows two patterns of distribution for the mitochondrial mutation: absence of the marker (clear area); presence of the marker (shaded area). (Courtesy of Gerald Shields)

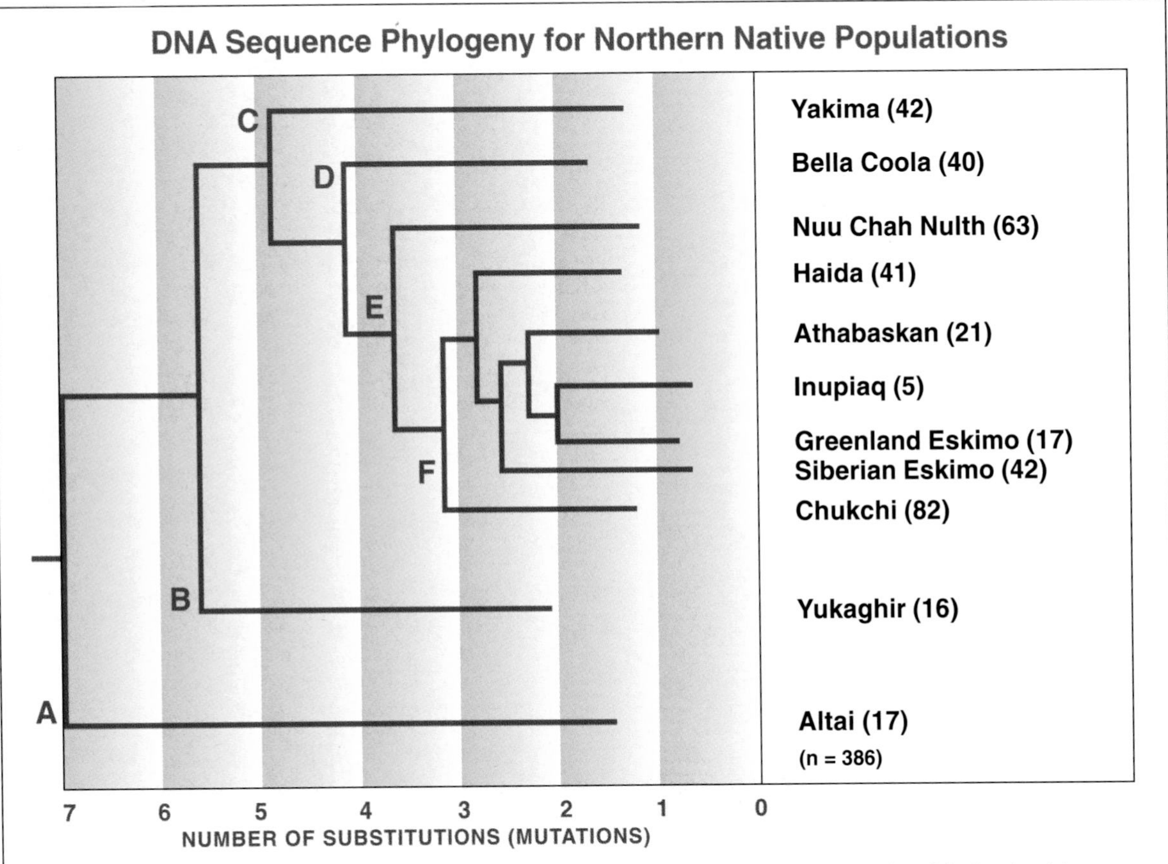

This tree of relationships shows the average genetic differences of any one group (i.e. Athabaskan) from another group based on the number of substitutions (mutations) that characterize the group. This tree of relationships is based on comparisons of 400 nucleotides of the mitochondrial control region from 386 individuals of Beringia and adjoining regions. The tree defines six major groupings (A-F). The Altai (A) and Yukaghir (B) populations are clearly separate from the other groups. The three Amerindian tribes defined by nodes C, D and E are also distinct and clearly separable from other tribes. However, the cluster of tribes defined by node F, which includes representatives of the Na-Dene language group (Haida and Athabaskan) and representatives of the Eskimo/Aleut language group (Greenlandic Eskimo, Inupiaq and Siberian Eskimo) shows no clear correlation between genetic diversity and language. (Source: *Gerald Shields; graphic by Kathy Doogan)*

(Athabaskans and Haida) do not separate out as unique tribal groups in analyses that depict genetic relationships among groups. Rather, Athabaskans and Haida are intermingled among Eskimo groups that speak Eskimo-Aleut. Possibly languages are changing more rapidly than genes and we should therefore be cautious about the use of linguistics to elucidate ancient relationships. It is possible that the genetic association between Eskimos and Na-Dene Indians has come about because of recent marriages between tribes of these groups. This seems unlikely because most of the DNA types that are shared between groups are ancient rather than recent.

We have also studied other genetic markers that support a close relationship between Eskimo-Aleuts and Na-Dene Indians. One of the genetic markers occurs between two mitochondrial genes and has been used to trace migrations out of Asia into the South Pacific. When this genetic length mutation is analyzed in groups of our study, two distributions are obvious. All indigenous northern groups from the Taymyr Peninsula of northcentral Russia across Siberia, Alaska, and the Arctic to East Greenland lack this length mutation. However, Altai, Buriats and Mongols of southern Siberia and all Amerind populations studied here possess the length mutation. Thus, it seems likely that lineages found among Amerind speakers are descended from a set of lineages that occurred in a genetically diverse set of early migrants. These early migrants, who introduced mitochondrial lineages with the length mutation, are presumed to have reached the New World first. Only after that event did populations that lacked the length mutation give rise to present-day Eskimo-Aleut and Na-Dene speakers.

So far four major groups of DNA sequence types have been observed in the New World. All Eskimo-Aleut and Na-Dene groups fall into only one of these groups. They therefore must be considered historically young and possibly have not existed long enough for evolutionary differentiation to have occurred in place. One cautionary note is important. All of our observations are based on the analysis of mitochondrial DNA that is genetically transmitted only by females. It is currently important for science to continue to develop detailed methods for the study of genes on the Y chromosome so that the genetics of migrations by males might also be studied. ■

Prehistoric Alaska on Display

Here are some places to see pieces of prehistoric Alaska:

■ **Alaska Resources Library and Bureau of Land Management Public Room, Federal Building, Anchorage.** Displays include fossil bones from *Edmontosaurus* dinosaurs, interpretive photos and information.

■ **Alaska State Museum, Juneau.** Permanent exhibits include artifacts from early Native people in Arctic and northwest coastal area before contact with outsiders.

■ **Alutiiq Cultural Center, Kodiak.** Numerous early cultural artifacts unearthed on Kodiak Island include relics from Ocean Bay culture 6,500 years ago, the Kachemak Tradition 2,000 to 3,000 years ago, and the Koniag culture 300 to 1,200 years ago. The center will relocate to the new Archaeological Repository Center scheduled to open in Kodiak in 1995.

■ **Anchorage Museum of History and Art.** Alaska Gallery displays include dioramas, artifacts, archaeological sequences and sites, information about Alaska Native groups before outside contact; hemispheric globe illustrates Bering Land Bridge theory of human migration into Alaska.

■ **Dorothy Page Museum, Wasilla.** Displays include actual fossil skull of nodosaurid dinosaur from Talkeetna Mountains with interpretive information.

■ **Southcentral Alaska Museum of Natural History, Eagle River.** New museum opened in summer 1994; exhibits include wildlife dioramas, display of invertebrates from the Talkeetna Mountains associated with nodosaurid dinosaur, and southcentral Alaska Tertiary plant fossils.

■ **The Pratt Museum, Homer.** Ten permanent exhibits dealing with regional prehistory include: Eskimo and Dena'ina Indian cultural artifacts; a display of the museum's 1987 Halibut Cove archaeological excavation with artifacts from the Kachemak tradition along with a scale model of a winter house; exhibits of Aleut artifacts; and a fossil exhibit including a rare perch, alder leaves, coal, wood and marine life such as mollusks.

■ **University of Alaska Museum, Fairbanks.** Fossil bones and skin impressions of *Edmontosaurus* hadrosaur dinosaurs; skull cast of *Pachyrhinosaurus* dinosaur; new dinosaur exhibit to open summer 1995; late Pleistocene "Blue Babe" bison exhibit; mummies of small Pleistocene mammals; bones of Pleistocene mammals including mastodon, lion, saber-toothed cat; tectonic plate exhibit about Alaska's formation; exhibit about glaciers; gold nugget display and discussion of mining methods that have uncovered many Pleistocene fossils.

■ **Wrangell Museum, Wrangell.** Petroglyph rock carvings by early coastal Indians 5,000 to 10,000 years ago; original totemic house poles from Chief Shakes Island represent precontact culture of the Stikine tribe of Tlingits.

■ **Outside Alaska.** Numerous museums exhibit prehistoric animals and early man in North America including materials from Alaska. Some of these are the American Museum of Natural History and the Smithsonian National Museum of the American Indian, both in New York; Smithsonian National Museum of Natural History, Washington D.C; Field Museum of Natural History, Chicago; Burke Museum, Seattle; The Museum of Paleontology, University of California, Berkeley; Denver Museum of Natural History, Denver; and the Royal Tyrrell Museum, Drumheller, Alberta, Canada.

Glossary

CALDERA: A large circular or oval crater created by a large volcanic eruption. The crater mostly results from subsidence as magma is removed from the underlying magma chamber; a less important cause is the material blasted out during eruption.

FAULTS: Faults are breaks in the Earth's crust where rock units and even soils slide past each other. In normal faults, rocks slide down the fault plane in the direction of gravity. In a thrust fault the fault plane lies at less than a 45 degree angle to horizontal and rocks on the upper side of the fault plane have been thrust over the lower rocks, counter to the direction of gravity. The motion on a strike-slip fault is typically horizontal, along the fault plane. In both normal and strike-slip faults, the fault plane is usually approximately vertically oriented. A fault system consists of multiple, braided or anastomosing faults of regional extent, where motion at any given time can be accommodated on one or more of the strands.

FLYSCH: Flysch is composed of thick deposits of alternating thin beds of sandstone and shale. Commonly gray to black, individual thin beds can be traced long distances, giving exposures a distinctive banded appearance. The deposits result from turbidity currents or surges of water carrying entrained sand and silt down the continental slope. The currents occur episodically, sometimes triggered by earthquakes. Much of the fill in oceanic trenches, such as the Aleutian Trench, is deposited by turbidity currents where the flysch deposits can be many thousands of feet thick. Flysch also can spread over the sea floor, particularly where rivers carry large amounts of sediment to the sea.

KULA PLATE: Kula is a Native American word meaning "all gone", used because originally it was thought the plate had been completely subducted. Later research found the Kula plate is not entirely gone and that it floors the deep ocean part of the Bering Sea. More recently, it was discovered that a small sliver of the Kula plate still exists south of the Aleutian trench at its far western end.

MAGMATIC ARC: Magma is the name applied to molten rock within the Earth's crust. Solidified at depth, it results in plutonic rocks, such as granite and diorite. Erupted at the surface from volcanoes, magma is the source of lava, ash and other volcanic products. An arcuate linear chain of volcanoes and underlying magma chambers typically result from the melting and assimilation of the descending plate in subduction zones. The arcuate chain of volcanoes is commonly referred to as a volcanic arc; if the underlying plutonic rocks or magma chambers are included, it is called a magmatic arc. Upon erosion of the volcanic rocks, the remains are often a linear group of granitic intrusions, the so-called root of the arc. The Aleutian Islands are a good example of a magmatic arc, visible is the volcanic arc.

MELANGE: Melange is the name applied to rocks that have been broken apart and the fragments chaotically mixed through tectonic processes. Typically, individual rock fragments may range in size between a few yards and a mile or so. Any rock type is possible in a melange; chert, volcanic rocks, and shale or siltstone from the sea-floor are common constituents. Near Anchorage, the Chugach Mountains are largely composed of melange; it is especially well-exposed along the Seward Highway between Potter Marsh and Falls Creek, between Bird Creek and Indian. Between Falls Creek and Indian, there is a transition between the melange and the sandstone and shale of the flysch deposits. Past Indian, the rocks the Seward Highway crosses are largely flysch deposits.

NIKOLAI GREENSTONE: Naming of rock units: Rock units, such as the Nikolai Greenstone or the Kanektok metamorphic complex, have formal or informal names, derived from the locality where they are best seen. A formal geologic name, always capitalized, refers to a rock unit having a specific set of geologic characteristics. The Nikolai Greenstone, a formal rock unit, is composed of a thick pile of altered subaerial basalt flows and submarine "pillow" lavas. The Kanektok metamorphic complex is an informally defined rock unit, consisting of loosely described metamorphic rocks in a general area.

OBDUCTION: This phenomenon occurs when an oceanic plate is thrust upon or overrides a continental plate. Because continental plates do not successfully subduct, obduction usually occurs only for limited distances over a continental plate. (See subduction.)

PALEOMAGNETIC SIGNATURE: When rocks form, they do so under the influence of the Earth's magnetic field at the time. Magnetic particles in the rocks tend to align themselves with the Earth's magnetic field, which at any given spot on the Earth depends largely on latitude. Hence, by measuring the alignment of the magnetic field of particles in a rock (the Paleomagnetic Signature), an estimate can be made of the latitude at which the rock formed (the paleolatitude). The paleolatitude for a rock can help constrain models describ-ing the past. Unfortunately, it is much more difficult to determine paleolongitude for a rock.

PERIGLACIAL: Originally used to describe conditions or processes on the margins of existing or former glaciers, its usage has expanded to include environments in which repeated freezing and thawing is an important factor. It also refers to features produced in these environments.

PERMAFROST: Permafrost is a property of any surficial deposit or bedrock in which a temperature below freezing has existed continuously for two years or more. The definition disregards the nature of the materials texture, water content and lithology. An "active" layer, which seasonally thaws, may occur over permafrost. Beneath the active layer, permafrost may extend from a few feet to thousands of feet below the surface. Permafrost underlies about one fifth of the world's land area.

REDOUBT: The 1989 eruption of Mount Redoubt was significant because of its destructive effects on the airline industry. A KLM Boeing 747 inadvertently flew through the ash cloud. All four of its engines stopped due to ash intake. Powerless, the plane lost more than 20,000 feet in altitude as the pilot tried to restart two engines, being successful only 1,500 feet above the Chugach Mountains. Inspection after an emergency landing in Anchorage showed the interiors of the engines were glazed and the plane received more $50 million damage. Engines and planes of other airlines were also damaged and the eruption disrupted travel as flights to and from Anchorage were canceled for the duration of the eruption.

RIFTING: Occurs when a plate is torn apart. Generally a result of rifting is the eruption of volcanic material between the two rifted parts. Active rifts on oceanic plates are called spreading centers; the mid-Atlantic Ridge is one.

STRUCTURE: Geologists use the word structure to refer to the form and internal character of rocks; it is smaller in scale than tectonic features. Structure usually indicates folding of sedimentary bedding or faulting and fracturing; but can also refer to columnar jointing in volcanic rocks resulting from cooling; or to banding, repetitive mineralogical changes in a rock. It is distinct from "texture" in that texture strictly refers to smallscale internal characteristics in a rock, such as the distribution of various sizes of grains.

SUBDUCTION: The collision of continental and oceanic crust results in subduction, where denser oceanic crust is overridden by less dense continental crust. However, as this occurs the continental rocks are somewhat deformed, and oceanic rocks may be scraped off and plastered against the lower side of the continental crust. At subduction zones, the descending oceanic crust partially melts and assimilates in the upper part of the Earth's mantle (typically about 90 to 95 miles deep) as it dives down within the Earth. The magmas that result rise through the Earth's crust, generally interacting with crustal rocks on the way, and finally solidifying in the crust as plutonic rocks or erupting as volcanoes. (See also magmatic arc.) Subduction zones are typically marked by deep trenches offshore of a linear chain of volcanoes. Although continental plates or terranes may start down a subduction zone, because of their buoyancy they do not reach the mantle and instead plug the subduction zone, halting the process.

TECTONIC: Tectonic refers to the larger-scale mechanical processes that affect rocks or structures that develop in rocks as a result of large-scale crustal movements. Processes include subduction, obduction and rifting. It also can refer to the nature of some earthquakes, as opposed to volcanically caused or man-made earthquakes. Volcanism is generally distinguished from tectonic effects, although they are interrelated.

TEPHRA: Tephra is a general term used for all blast-related products of a volcano. It includes ash, block-and-ash fall deposits, and other fragmental eruptive products. It does not include lava flows. Volcanoes generally seen in Alaska produce large volumes of tephra as compared to volcanoes in Hawaii.

TERRANE: Terrane is used in a number of senses in geology. Spelled this way, it refers to an area of regional extent bounded by faults and having a distinct geologic history relative to surrounding terranes. Often, part of the definition of a terrane includes reference to a unique fossil fauna and flora. In some usages, terranes are also considered discrete fragments of oceanic or continental material added to a continental mass at an active margin. Because the definition of terranes is an interpretive process, there can be considerable controversy resulting from the definition of a terrane depending on the individual scientist. Spelled terrain, it means a tract or region of the Earth's surface considered as a physical feature. Terrain has meaning in the sense of form rather than content, such as when used as terrane.

TIME: The statement "many hundreds of million years" is often made by geologists; it is difficult for much of the public to understand what this really means. Often the analogy is made that man's time on Earth corresponds to the last few seconds before midnight in the "day" of the Earth. However, this obscures the fact that we are nowhere near the end of the "Earth's day" and the processes acting on the Earth will continue to act for long into the future. Another perspective on the concept of time is that if we were to begin a walk to the sun, the first mile in the distance is equivalent to 50 years of the time back to the beginning of the Earth. Taking a rest break on the moon would be equivalent to having gone back in time 120 million years, to the age of dinosaurs. In either case, there is a very long way to go. Yet, some of the Earth's tectonic plates, moving 2.5 inches per year, will move more than 4,700 miles in 120 million years. To man, this is a generally imperceptible rate of motion, yet because of the very long times involved, Alaska has changed much and will continue to change. ■

Bibliography

Anderson, Elaine. "Who's Who in the Pleistocene: A Mammalian Bestiary," *Quaternary Extinctions*. Tucson, Ariz.: University of Arizona Press, 1984.

Anthony, L.M. and A. T. Tunley. *Introductory Geography and Geology of Alaska*. Anchorage: Polar Publishing, 1976.

Bakker, Robert T. *The Dinosaur Heresies*. New York: Kensington Publishing Co., 1993.

Benton, Michael. *Dinosaur Factfinder*. New York: Kingfisher Books, 1992.

Brouwers, Elizabeth, William Clemens, Robert Spicer, Thomas Ager, et al. "Dinosaurs on the North Slope, Alaska: High Latitude, Latest Cretaceous Environments," *Science*,Vol. 237. Washington D.C.: American Association for the Advancement of Science, Sept. 1987.

Burk, C.A. *Geology of the Alaska Peninsula—Island Arc and Continental Margin*. New York: Geological Society of America Memoir 99, 1965.

Casadevall, T.J. *First International Symposium on Volcanic Ash and Aviation Safety, Program and Abstracts, Seattle, Wa, July 8-12, 1991*. U.S. Geological Survey Circular 1065, 1991.

Csejtey, Bela, Jr. and D.R. St. Aubin. "Evidence for northwestward thrusting of the Talkeetna superterrane, and its regional significance," in Albert, N.R.D. and Travis Hudson, editors. *The United States Geological Survey in Alaska: Accomplishments during 1979*. U.S. Geological Survey Circular 823-B. 1981.

___, D.P. Cox, R.C. Evarts, G.D. Stricker and H.L. Foster. "The Cenozoic Denali Fault System and the Cretaceous Accretionary Development of Southern Alaska." Journal of Geophysical Research, Vol. 87, No. B5. 1982.

Clemens, William A. "Continental Vertebrates from the Late Cretaceous of the North Slope, Alaska." in Thurston, Dennis K. and Kazuya Fujita, editors.*1992 Proceedings, International Conference on Arctic Margins*. Anchorage: Alaska Region, U.S. Minerals Management Service, 1994.

___, and L. Gayle Nelms. "Paleoecological implications of Alaskan terrestrial vertebrate fauna in latest Cretaceous time at high paleolatitudes," *Geology*, Vol. 21. Boulder, Colo.: Geological Society of America, Inc., June 1993.

Coats, R.R. "Volcanic Activity in the Aleutian Arc" in *Contributions to General Geology*. U.S. Geological Survey Bulletin 974-B, 1950.

Crowley, T.J. and G.R. North. *Paleoclimatology*. New York: Oxford University Press, 1991.

Currie, Philip J. "Long-distance Dinosaurs," *Natural History*. New York: American Museum of Natural History, June 1989.

Detterman, R.L., J.E. Case, F.H. Wilson and M.E. Yount. *Geologic Map of the Ugashik, Bristol Bay, and Western Part of Karluk Quadrangles, Alaska*. U.S. Geological Survey Miscellaneous Investigations Series Map I-1685. 1987.

___, J.W. Miller, F.H. Wilson and M.E. Yount. *Stratigraphic Framework of the Alaska Peninsula*. U.S. Geological Survey Bulletin 1969. In press.

___, T.P. Miller, M.E. Yount and F.H. Wilson. *Geologic Map of the Chignik and Sutwik Island Quadrangles, Alaska*. U.S. Geological Survey Miscellaneous Investigations Series Map I-1229.

___, and B.L. Reed. *Stratigraphy, Structure, and Economic Geology of the Iliamna Quadrangle, Alaska*. U.S. Geological Survey Bulletin 1368-B, 1980.

Drewes, Harold, G.D. Fraser, G.L. Snyder and H.F. Barnett, Jr. "Geology of Unalaska Island and Adjacent Insular Shelf, Aleutian Islands, Alaska," in *Investigations of Alaska Volcanoes*. U.S. Geological Survey Bulletin 1028-S. 1961.

Eakins, G.R. *A Petrified Forest on Unga Island, Alaska*. Alaska Division of Mines and Geology Special Report 3, 1970.

Ewer, R.F. *The Carnivores*. London: Weidenfeld and Nicolson, 1973.

Ferrians, O.J., Jr., Reuben Kachadoorian and G.W. Greene. *Permafrost and Related Engineering Problems in Alaska*. U.S. Geological Survey Professional Paper 678, 1969.

Foster, H.L., compiler. *Geologic Map of the Eastern Yukon-Tanana Region, Alaska*. U.S. Geological Survey Open-file Report 92-313. 1992.

Frakes, L.A., J.E. Francis and J.I. Syktus. *Climate Modes of the Phanerozoic, the History of the Earth's Climate Over the Past 600 Million Years*. Cambridge: Cambridge University Press, 1992.

Gangloff, Roland. "Polar Dinosaurs," *Environment West*. Fairbanks: University of Alaska Museum, 1990.

___. "*Edmontonia* Sp., The First Record of an Ankylosaur from Alaska." *Journal of Vertebrate Paleontology*. Lincoln, Neb.: University of Nebraska, 1994.

Gates, Olcott, H.A. Powers and R.E. Wilcox. "Geology of the Near Islands, Alaska" with a section on "Surficial Geology of the Near Islands" by J.P. Schafer in Investigations of Alaska Volcanoes. U.S. Geological Survey Bulletin 1028-U. 1971.

Giddings, James L. *Ancient Men of the Arctic*. New York: Alfred A. Knopf, 1967.

___ and Douglas D. Anderson. *Beachridge Archaeology of Cape Krusenstern*. National Park Service Publications in Archaeology 20. Washington D.C.: U.S. Department of the Interior, 1986.

Gore, Rick. "Dinosaurs." *National Geographic*, Vol. 183, No. 1. Washington, D.C.: National Geographic Society, January 1993.

Greenberg, J.H. *Languages in the Americas*. Stanford, Calif.: Stanford University Press, 1987.

Guthrie, R. D. *Frozen Fauna of the Mammoth Steppe, The Story of Blue Babe*. Chicago: University of Chicago Press, 1990.

___. "Paleoecology of a late Pleistocene small mammal community from interior Alaska," *ARCTIC*. Vol. 22, pp. 213-224. Calgary, Alberta: University of Calgary, 1968.

___. "Paleoecology of the Large Mammal Community in Interior Alaska During the Late Pleistocene," *American Midland Naturalist*. Vol. 79, No. 2, pp. 346-363. Notre Dame, Ind.: University of Notre Dame, 1968.

___. "Bison and man in North America," *Canadian Journal of Anthropology*. Vol. 1, pp. 55-73. Edmonton, Alberta: University of Alberta, Dept. of Anthropology, 1980.

Hamilton, Thomas D., Katherine M. Reed, Robert M. Thorson, editors. *Glaciation in Alaska, The Geologic Record*. Anchorage: Alaska Geological Society, 1986.

Harington, C.H. "Quaternary Vertebrate Faunas of Canada and Alaska and Their Suggested Chronological Sequence," *Syllogeus*. No. 15. Ottawa, Ontario: National Museums of Canada, National Museum of Natural Sciences, 1978.

Harris, A.H. *Late Pleistocene Vertebrate Paleoecology of the West*. Austin, Texas: University of Texas Press, 1985.

Harritt, Roger K. *Eskimo Prehistory on the Seward Peninsula, Alaska*. Resources Report NPS/ARO RCR/CRR-93/21. Washington D.C.: National Park Service, U.S. Department of the Interior, 1994.

Hays, W.W., editor. *Facing Geologic and Hydrologic Hazards, Earth Science Considerations*. U.S. Geological Survey Professional Paper 1240-B, 1981.

Henning, R.A., Barbara Olds and Penny Rennick, editors. *A Photographic Geography of Alaska*. Anchorage: Alaska Geographic Society, 1980.

Hillhouse, J.W. "Accretion of Southern Alaska." in *Tectonophysics*. Vol. 139. 1987.

Hopkins, David M., editor. *The Bering Land Bridge*. Stanford, Calif.: Stanford University Press, 1967.

___, J.V. Matthews, Jr., C.E. Schweger, S.B. Young, editors. *Paleoecology of Berengia*. New York: Academic Press, 1982.

Jones, D.L., N.J. Silberling and John Hillhouse. "Wrangellia—A Displaced Terrane in Northwestern North America." *Canadian Journal of Earth Science*, Vol. 14, 1977.

Kennedy, G.C. and H.H. Waldron. "Geology of Pavlof Volcano and Vicinity, Alaska" in *Investigations of Alaska Volcanoes*. U.S. Geological Survey Bulletin 1028-A, 1955.

Kontrimavichus, V.L., editor. *Beringia in the Cenozoic Era*. Rotterdam: A.A. Balkema, 1985.

Kurten, Bjorn. *Pleistocene Mammals of Europe*. London: The World Naturalist Publishers, 1965.

___ and Elaine Anderson. *Pleistocene Mammals of North America*. New York, N.Y.: Columbia University Press, 1980.

Larsen, Helge. "Trail Creek: Final Report on the Excavation of Two Caves on the Seward Peninsula, Alaska." Copenhagen: *Acta Arctica* Vol. 15, 1968.

Lessem, Don. *Kings of Creation*. New York: Simon and Schuster, 1992.

Mertie, J.B. *The Yukon-Tanana Region, Alaska*. U.S. Geological Survey Bulletin 872. 1937.

Molnia, Bruce. *Alaska's Glaciers*. Anchorage: Alaska Geographic Society, 1982.

Morell, Virginia. "How Lethal was the K-T Impact?" *Science*, Vol. 261. Washington, D.C.: American Association for the Advancement of Science, Sept. 1993.

Mull, C.G. and I.L. Trailleur. "A Skeleton in Triassic Rocks in the Brooks Range Foothills," *ARCTIC*, Vol. 26, No. 1. Calgary, Alberta: University of Calgary, March 1973.

Nuhfer, E.B., R.J. Proctor, P.H. Moser and others. *The Citizens' Guide to Geologic Hazards*. American Institute of Professional Geologists, 1993.

Oliver, J.E. *Climate and Man's Environment, an Introduction to Applied Climatology*. New York: John Wiley and Sons, Inc., 1973.

Olson, Wallace M. *The Alaska Travel Journal of Archibald Menzies, 1793-1794*. Fairbanks: University of Alaska Press, 1993.

___. *The Tlingit: An introduction to their culture and history*. Juneau: Heritage Research, 1993.

Price, L.W. *The Periglacial Environment, Permafrost, and Man*. Association of American Geographers, Commission on College Geography, Resource Paper No. 14., 1972.

Ray, L.L. *Permafrost*. U.S. Geological Survey General Interest Publication, 1993 printing.

Reid, Monty. *Last Great Dinosaurs, An Illustrated Guide to Alberta's Dinosaurs*. Red Deer, Alberta: Red Deer College Press, 1990.

Rennick, Penny, editor. *Alaska's Oil/Gas & Minerals Industry*. Anchorage: Alaska Geographic Society, 1982.

—, *Alaska's Volcanoes*. Anchorage: Alaska Geographic Society, 1991.

Riehle, J.R., R.L. Detterman, J.W. Miller and M.E. Yount. Geologic Map of the Mt. Katmai Quadrangle and Portions of the Afognak and Naknek Quadrangles, Alaska. U.S. Geological Survey Miscellaneous Investigations Series Map. In press.

Rotberg, R.I. and T.K. Rabb, editors. *Climate and History, Studies in Interdisciplinary History*. Princeton, N.J.: Princeton University Press, 1981.

Schaaf, Jeanne M. *The Bering Land Bridge National Preserve: an Archaeological Survey*. Vols. I and II. National Park Service Resource/Research Management Report AR-14. Washington D.C.: U.S. Department of the Interior, 1988.

___. "Cooperative Research and Resource Management in Protected Areas: A Case Study in the Proposed Beringian Heritage International Park." In *Proceedings of Human Ecology and Global Climate Change: The Role of Parks and Protected Areas*. Washington D.C.: Taylor and Francis. In press.

Shields, G.F., K. Hecker, M. Voevoda and J.K. Reed. "Absence of the Asian-specific region V mitochondrial marker in native Beringians." *American Journal of Human Genetics*, 50:758-765. 1992.

___, A. Schmiechen, B. Frazier, A. Redd, M.I. Voeevoda, J.K. Reed and R.H. Ward. "mt DNA Sequences Suggest a Recent Evolutionary Divergence for Beringian and Northern North American Populations." *American Journal of Human Genetics.*, 53:549-562, 1993.

Spicer, Robert A, Judith Totman Parrish, Paul R. Grant. "Evolution of vegetation and coal-forming environments in the Late Cretaceous of the North Slope of Alaska," *Controls on the Distribution and Quality of Cretaceous Coals,*. Special Paper 267. Boulder, Colo.: The Geological Society of America, Inc., 1992.

Stoneking, M. and A.C. Wilson. "Mitochondrial DNA." In: Hill, A., and S. Serjeantson, editors. *The colonization of the Pacific: a genetic trail*. Oxford, England: Oxford University Press, 1989.

Szathmary, E.J.E. "mtDNA and the Peopling of the Americas." *American Journal of Human Genetics*, 53:793-799, 1993.

Vinson, Dale M. "Taphonomic Analysis of Faunal Remains from Trail Creek Caves, Seward Peninsula, Alaska." Unpublished Master Thesis. Fairbanks: University of Alaska, Dec. 1993.

Wahrhaftig, Clyde, D.L. Turner, F.R. Weber and T.E. Smith. "Nature and Timing of Movement of Hines Creek Strand of Denali Fault System, Alaska." *Geology*, Vol. 3, No. 8, 1975.

Weller, Gunter and S.A. Bowling, editors. *Climate of the Arctic*. Fairbanks: Geophysical Institute, University of Alaska, 1975.

Wigley, T.M.L., M.J. Ingram and G. Farmer. *Climate and History, Studies in Past Climates and Their Impact on Man*. Cambridge: Cambridge University Press, 1981.

Wilson, F.H., R.L. Detterman and J.E. Case. *The Alaska Peninsula Terrane: A Definition*. U.S. Geological Survey Open-File Report 85-450, 1985.

___, R.L. Detterman, J.W. Miller and J.E. Case. *Geologic Map of the Port Moller, Stepovak Bay, and Simeonof Island Quadrangles, Alaska*. U.S. Geological Survey Miscellaneous Investigations Series Map I-2272. In press.

Index

ALASKA GEOGRAPHIC® Back issues

The North Slope, Vol. 1, No. 1. Out of print.

One Man's Wilderness, Vol. 1, No. 2. Out of print.

Admiralty...Island in Contention, Vol. 1, No. 3. $7.50.

Fisheries of the North Pacific, Vol. 1, No. 4. Out of print.

Alaska-Yukon Wild Flowers Guide, Vol. 2, No. 1. Out of print.

Richard Harrington's Yukon, Vol. 2, No. 2. Out of print.

Prince William Sound, Vol. 2, No. 3. Out of print.

Yakutat: The Turbulent Crescent, Vol. 2, No. 4. Out of print.

Glacier Bay: Old Ice, New Land, Vol. 3, No. 1. Out of print.

The Land: Eye of the Storm, Vol. 3, No. 2. Out of print.

Richard Harrington's Antarctic, Vol. 3, No. 3. $12.95.

The Silver Years, Vol. 3, No. 4. $17.95.

Alaska's Volcanoes: Northern Link In the Ring of Fire, Vol. 4, No. 1. Out of print.

The Brooks Range, Vol. 4, No. 2. Out of print.

STATEMENT OF OWNERSHIP, MANAGEMENT & CIRCULATION

ALASKA GEOGRAPHIC® is a quarterly publication, home office at P.O. Box 93370, Anchorage, AK 99509. Editor is Penny Rennick. Publisher and owner is The Alaska Geographic Society, a non-profit Alaska organization, P.O. Box 93370, Anchorage, AK 99509. *ALASKA GEOGRAPHIC®* has a membership of 6,117.

Total number of copies		**9,221**
Paid and/or requested circulation		
Sales through dealers, etc.	50	
Mail subscription	6,117	
Total paid and/or requested circulation	6,167	
Free distribution	50	
Total distribution	6,217	
Copies not distributed (office use, returns, etc.)	3,004	
TOTAL		**9,221**

I certify that the statement above is correct and complete.

—Vickie Staples
Circulation/Database Manager

Kodiak: Island of Change, Vol. 4, No. 3. Out of print.

Wilderness Proposals, Vol. 4, No. 4. Out of print.

Cook Inlet Country, Vol. 5, No. 1. Out of print.

Southeast: Alaska's Panhandle, Vol. 5, No. 2. Out of print.

Bristol Bay Basin, Vol. 5, No. 3. Out of print.

Alaska Whales and Whaling, Vol. 5, No. 4. $19.95.

Yukon-Kuskokwim Delta, Vol. 6, No. 1. Out of print.

Aurora Borealis, Vol. 6, No. 2. $19.95.

Alaska's Native People, Vol. 6, No. 3. $24.95.

The Stikine River, Vol. 6, No. 4. $15.95.

Alaska's Great Interior, Vol. 7, No. 1. $17.95.

Photographic Geography of Alaska, Vol. 7, No. 2. Out of print.

The Aleutians, Vol. 7, No. 3. Out of print.

Klondike Lost, Vol. 7, No. 4. Out of print.

Wrangell-Saint Elias, Vol. 8, No. 1. Out of print.

Alaska Mammals, Vol. 8, No. 2. Out of print.

The Kotzebue Basin, Vol. 8, No. 3. Out of print.

Alaska National Interest Lands, Vol. 8, No. 4. $17.95.

Alaska's Glaciers, Vol. 9, No. 1. Revised 1993. $19.95.

Sitka and Its Ocean/Island World, Vol. 9, No. 2. Out of print.

Islands of the Seals: The Pribilofs, Vol. 9, No. 3. $15.95.

Alaska's Oil/Gas & Minerals Industry, Vol. 9, No. 4. $15.95.

Adventure Roads North, Vol. 10, No. 1. $17.95.

Anchorage and the Cook Inlet Basin, Vol. 10, No. 2. $17.95.

Alaska's Salmon Fisheries, Vol. 10, No. 3. $15.95.

Up the Koyukuk, Vol. 10, No. 4. $17.95.

Nome: City of the Golden Beaches, Vol. 11, No. 1. $15.95.

Alaska's Farms and Gardens, Vol. 11, No. 2. $15.95.

Chilkat River Valley, Vol. 11, No. 3. $15.95.

Alaska Steam, Vol. 11, No. 4. $15.95.

Northwest Territories, Vol. 12, No. 1. $17.95.

Alaska's Forest Resources, Vol. 12, No. 2. $16.95.

Alaska Native Arts and Crafts, Vol. 12, No. 3. $19.95.

Our Arctic Year, Vol. 12, No. 4. $15.95.

Where Mountains Meet the Sea: Alaska's Gulf Coast, Vol. 13, No. 1. $17.95.

Backcountry Alaska, Vol. 13, No. 2. $17.95.

British Columbia's Coast, Vol. 13, No. 3. $17.95.

Lake Clark/Lake Iliamna Country, Vol. 13, No. 4. Out of print.

Dogs of the North, Vol. 14, No. 1. $17.95.

South/Southeast Alaska, Vol. 14, No. 2. Out of print.

Alaska's Seward Peninsula, Vol. 14, No. 3. $15.95.

The Upper Yukon Basin, Vol. 14, No. 4. $17.95.

Glacier Bay: Icy Wilderness, Vol. 15, No. 1. Out of print.

Dawson City, Vol. 15, No. 2. $15.95.

Denali, Vol. 15, No. 3. $16.95. Out of print.

The Kuskokwim River, Vol. 15, No. 4. $17.95.

Katmai Country, Vol. 16, No. 1. $17.95.

North Slope Now, Vol. 16, No. 2. $15.95.

The Tanana Basin, Vol. 16, No. 3. $17.95.

The Copper Trail, Vol. 16, No. 4. $17.95.

The Nushagak Basin, Vol. 17, No. 1. $17.95.

Juneau, Vol. 17, No. 2. Out of print.

The Middle Yukon River, Vol. 17, No. 3. $17.95.

The Lower Yukon River, Vol. 17, No. 4. $17.95.

Alaska's Weather, Vol. 18, No. 1. $17.95.

Alaska's Volcanoes, Vol. 18, No. 2. $17.95.

Admiralty Island: Fortress of the Bears, Vol. 18, No. 3. $17.95.

Unalaska/Dutch Harbor, Vol. 18, No. 4. $17.95.

Skagway: A Legacy of Gold, Vol. 19, No. 1. $18.95.

ALASKA: The Great Land, Vol. 19, No. 2. $18.95.

Kodiak, Vol. 19, No. 3. $18.95.

Alaska's Railroads, Vol. 19, No. 4. $18.95.

Prince William Sound, Vol. 20, No. 1. $18.95.

Southeast Alaska, Vol. 20, No. 2. $19.95.

Arctic National Wildlife Refuge, Vol. 20, No. 3. $18.95.

Alaska's Bears, Vol. 20, No. 4. $18.95.

The Alaska Peninsula, Vol. 21, No. 1. $19.95.

The Kenai Peninsula, Vol. 21, No. 2. $19.95.

People of Alaska, Vol. 21, No. 3. $19.95.

Prehistoric Alaska, Vol. 21, No. 4. $19.95.

ALL PRICES SUBJECT TO CHANGE

Your $39 membership in The Alaska Geographic Society includes four subsequent issues of *ALASKA GEOGRAPHIC®*, the Society's official quarterly. Please add $10 per year for non-U.S. memberships.

Additional membership information and free catalog are available on request. Single *ALASKA GEOGRAPHIC®* back issues are also available. When ordering, please make payments in U.S. funds and add $2.00 postage/handling per copy for Book Rate; $4.00 per copy for Priority Mail. Inquire for non-U.S. postage rates. To order back issues send your check or money order or credit card information (including expiration date and daytime phone number) and volumes desired to:

The Alaska Geographic Society

P.O. Box 93370
Anchorage, AK 99509-3370
Phone (907) 562-0164; Fax (907) 562-0479

NEXT ISSUE: *Fairbanks*, Vol. 22, No. 1. At the center of Alaska's golden heart lies the state's second largest city, Fairbanks. From its origin as a mining supply depot on the Chena River, Fairbanks has grown into the service and supply center for Alaska's Interior and North. Education, military activities and tourism complement this role and keep the golden heart beating steadily. To members 1995, with index. Price $19.95.